Compendium of Flowering Potted Plant Diseases

by

Margery L. Daughtrey
Cornell University
Long Island Horticultural Research Laboratory
Riverhead, New York

Robert L. Wick
University of Massachusetts
Amherst

Joseph L. Peterson
Rutgers University
New Brunswick, New Jersey

APS PRESS
The American Phytopathological Society

Cover photographs courtesy Margery Daughtrey and Daniel Gilrein

Library of Congress Catalog Card Number: 95-78639
International Standard Book Number: 0-89054-202-3

Printed in the United States of America

The American Phytopathological Society
3340 Pilot Knob Road
St. Paul, MN 55121-2097, USA

Preface

This compendium brings together information on the diseases of an important floriculture crop group, the greenhouse-grown, flowering potted plants. In the 1993 U.S. Department of Agriculture Floriculture Crops Summary, flowering potted crops in the United States (including potted geraniums) were valued at $808 million wholesale. Poinsettias and geraniums are currently the dominant flowering potted-plant crops in the United States. A greater diversity and volume of flowering potted plants is marketed in Europe than in the United States, where sales are closely associated with major holidays.

Disease prevention and management are important concerns of potted-plant growers. Plant diseases cause crop losses directly through plant mortality and frequently cause aesthetic injury that reduces crop quality and value.

In the course of their production from seed or cuttings to finished hanging basket or pot, plants are often moved from one greenhouse operation to another, sometimes even internationally. This practice creates opportunities for the introduction of plant pathogens from one greenhouse operation to another. Greenhouse-to-greenhouse exchange of plants also facilitates the exchange of pesticide-resistant diseases and insects throughout the greenhouse industry worldwide.

Because many different potted-plant species are typically produced within one operation, it is especially challenging for the grower to be thoroughly informed of all the potential disease problems. Regular monitoring should be conducted with full awareness of the diseases important to each crop. Anticipating likely problems also allows growers to take preventive cultural approaches to disease management. Prevention strategies are much more effective than responsive treatments after symptoms have developed. Once symptoms are apparent, prompt and accurate diagnosis is essential so that appropriate disease-control measures can be instituted without delay.

It is the authors' hope that this compendium will provide a helpful reference to growers for both diagnosis and implementation of better preventive and responsive disease management programs for their crops. The information reported has been extracted from the available scientific literature and the authors' experience. Growers will find the Host Range and Epidemiology and Management sections within each disease discussion to be most helpful.

The management discussions will in some instances appear incomplete, reflecting a real lack of information regarding the control of many of these diseases. Fungicides are used to control many of the diseases described in this compendium. They are mentioned within these pages for information purposes only. Legal restrictions and regulations of fungicide use vary among countries, and regulations within any country are subject to change over time. Therefore, mention of pesticides in this compendium must not be taken as a recommendation or evidence that the pesticides are registered for use on any given crop. Under no circumstances should a chemical control method be used unless the chemical is specifically registered for the intended method of application on the crop. Consult local pesticide labeling regulations for recommendations and restrictions. In the United States, the Cooperative Extension Service should be consulted.

Descriptions of pathogens (in the Causal Organism sections) are provided in sufficient detail to assist professional diagnosticians in making accurate diagnoses. The information in the Causal Organism and Host Range and Epidemiology sections will assist pest-management specialists in the design of programs that take into account each pathogen's life cycle, methods of dispersal, and responses to environmental factors.

Not every crop grown as a flowering potted plant has been addressed within this volume. Notable omissions include roses and azaleas (addressed in APS Press's *Compendium of Rose Diseases* and *Compendium of Rhododendron and Azalea Diseases,* respectively) and chrysanthemums (to be addressed in a forthcoming volume from APS Press). Bedding plants marketed primarily in packs or flats have also been excluded from this volume, although crops such as wax begonias (*Begonia* Semperflorens-Cultorum hybrids) and impatiens (*Impatiens wallerana*), usually produced in flats, are included because of their close taxonomic similarity to the important potted crops Rieger and nonstop begonias and New Guinea impatiens.

We have not attempted to cover greenhouse crop species primarily produced as cut flowers (such as carnations and alstroemeria), cacti, or orchids. We leave these topics to the authors of future compendia.

For additional reading on the diseases of flowering potted plants, an excellent comprehensive reference is *Diseases of Floral Crops,* volumes 1 and 2, edited by D. L. Strider (Praeger Press). *Diseases of Greenhouse Plants* by J. T. Fletcher (Longman Group, Ltd.) supplies valuable information on the management of diseases of vegetable and ornamental crops. *Managing Diseases in Greenhouse Crops* by William R. Jarvis (APS Press) provides a thorough review of the science of disease management in greenhouses.

In order to avoid repetition, the diseases in this compendium have been presented by pathogen rather than by host. The Quick Guide to the Flowering Potted Plants and their Infectious Diseases, which lists all of the diseases mentioned in the text for each of the crop genera, is for readers who wish to study all the diseases of a particular flowering potted plant. Those wishing information on a particular pathogen will find the table of contents helpful, and the index at the end of the volume allows readers to find diseases by pathogen or common name.

The authors sincerely thank the many people who assisted them in the preparation of this compendium. The following individuals provided invaluable help and critical comments on portions of the manuscript: Ron Adams, Ball Seed Company, West Chicago, IL; Thomas Boyle, University of Massachusetts, Amherst; Edward Brown, II, University of Georgia, Athens; A. R. Chase, Chase Research Gardens, Thousand Oaks, CA; Douglas Cox, University of Massachusetts, Amherst; Robert E. Davis, U.S. Department of Agriculture,

Beltsville, MD; Ralph Freeman, Cornell Cooperative Extension of Suffolk County, Riverhead, NY; Daniel Gilrein, Cornell Cooperative Extension, Riverhead, NY; Ann Brooks Gould, Rutgers, The State University of New Jersey, Cook College, New Brunswick; Susan Han, University of Massachusetts, Amherst; Mary K. Hausbeck, Michigan State University, East Lansing; Will Healy, Ball Seed Company, West Chicago, IL; Ronald K. Jones, North Carolina State University, Raleigh; Roger Lawson, Florist and Nursery Crops Laboratory, U.S. Department of Agriculture, Beltsville, MD; Kurt Leonard, Cereal Rust Laboratory, University of Minnesota, St. Paul; Richard Lindquist, The Ohio State University, Wooster; William Manning, University of Massachusetts, Amherst; James A. Matteoni, Kwantlen College, Surrey, British Columbia, Canada; Gary Moorman, Pennsylvania State University, State College; James Moyer, North Carolina State University, Raleigh; Steven Nameth, The Ohio State University, Columbus; Amy Rossman, Mycology Laboratory, Beltsville Agricultural Research Center, Beltsville, MD; John P. Sanderson, Cornell University, Ithaca, NY; and Andrew Senesac, Cornell Cooperative Extension, Riverhead, NY.

Others assisted with research and manuscript preparation: Carole Brush, Cornell University, Riverhead, NY; Bruce Clarke, Rutgers, The State University of New Jersey, Cook College, New Brunswick; Richard Craig, Pennsylvania State University, State College; Patricia Haviland, University of Massachusetts, Amherst; Kent Loeffler, Cornell University, Ithaca, NY; Maria Macksel, Cornell University, Riverhead, NY; Sandi Mulvaney, Cornell University, Riverhead, NY; Oya Rieger, Mann Library, Cornell University, Ithaca, NY; Pat Pogers, Rutgers, The State University of New Jersey, Cook College, New Brunswick; Leanne Pundt, Connecticut Cooperative Extension, Haddam; and the editorial staff of APS Press.

We are especially grateful to those who shared their slides of pathogens or disease symptoms: Wayne R. Allen, Agriculture Canada, Vineland Station, Ontario; Minoru Aragaki, University of Hawaii, Manoa; Larry Barnes, Texas A&M University, College Station; Karen Bech, National Institute of Plant Pathology, Lyngby, Denmark; Inge Bouwen, Research Institute for Plant Protection, Wageningen, The Netherlands; Timothy Bowyer, Norcross, GA; Peggy Brannigan, Floral and Nursery Crops Research Unit, U.S. Department of Agriculture, Beltsville, MD; Gail Celio, Michigan State University, East Lansing; A. R. Chase, Chase Research Gardens, Thousand Oaks, CA; Douglas Cox, University of Massachusetts, Amherst; Thomas Cresswell, North Carolina State University, Raleigh; Spencer H. Davis, North Brunswick, NJ; Ethel M. Dutky, University of Maryland, College Park; Wilhelm Elsner, Elsner pac Jungpflanzen, Dresden, Germany; Arthur W. Engelhard, University of Florida, Bradenton; John Fletcher, ADAS Wye, Kent, England; Daniel Gilrein, Cornell Cooperative Extension, Riverhead, NY; Mark Gleason, Iowa State University, Ames; Garry Grüber, Kientzler KG, Gensingen, Germany; Mary Ann Hansen, Virginia Polytechnic Institute and State University, Blacksburg; Mary Hausbeck, Michigan State University, East Lansing; Sister Mary Francis Heimann, University of Wisconsin, Madison; William Jarvis, Agriculture Canada, Harrow, Ontario; Ronald K. Jones, North Carolina State University, Raleigh; Diane Karasevicz, Cornell University, Ithaca, NY; Michael J. Klopmeyer, Ball FloraPlant, West Chicago, IL; Conrad Krass, California Department of Food and Agriculture, Sacramento; Seong Hwan Kim, Pennsylvania Department of Agriculture, Bureau of Plant Industry, Harrisburg; Roger Lawson, Floral and Nursery Crops Research Unit, U.S. Department of Agriculture, Beltsville, MD; Bent Løschenkohl, Danish Institute of Plant and Soil Science, Lyngby, Denmark; Maria Macksel, Cornell University, Riverhead, NY; William J. Manning, University of Massachusetts, Amherst; James Matteoni, Kwantlen College, Surrey, British Columbia, Canada; John J. McRitchie, Florida Department of Agriculture, Gainesville; John J. Mishanec, Cornell Cooperative Extension, Voorheesville, NY; Steven Nameth, The Ohio State University, Columbus; Michael P. Parrella, University of California, Davis; George Philley, Texas A&M Research and Extension Center, Overton; Luellen Pierce, University of California, Berkeley; Jay Pscheidt, Oregon State University, Corvallis; Leanne S. Pundt, University of Connecticut, Haddam; John P. Sanderson, Cornell University, Ithaca, NY; Andrew Senesac, Cornell Cooperative Extension, Riverhead, NY; Malcolm Shurtleff, University of Illinois, Urbana; Gary W. Simone, University of Florida, Gainesville; Ward C. Stienstra, University of Minnesota, St. Paul; Janice Uchida, University of Hawaii, Manoa; and Ruth A. Welliver, Pennsylvania Department of Agriculture, Bureau of Plant Industry, Harrisburg.

The authors express their deepest appreciation and gratitude to Robert J. Kent, Lynn F. Wick, and Karen R. Peterson for their unflagging support through the long duration of this project.

Margery L. Daughtrey
Robert L. Wick
Joseph L. Peterson

Contents

Quick Guide to the Flowering Potted Plants and their Infectious Diseases

Table 1 provides a list of the scientific and common names of each plant discussed in this compendium. The following is a brief discussion of the history and cultivation of each plant and a list of the diseases and insects to which it is susceptible.

Anemone coronaria (Ranunculaceae), Anemone

Anemone coronaria L., native to the northern coast of the Mediterranean, has in the past been grown under cool conditions from tuberous rhizomes as a greenhouse cut flower. Recent F_1 hybrids (including the Mona Lisa series) allow production from seed, making the anemone more adaptable as a potted plant and decreasing the likelihood of disease problems. The new hybrids provide a wide color range, including wine, pink, white, blue, orchid, red, and bicolors. Aphids, whiteflies, leafminers, and thrips are the major insect problems for anemone. The major disease problems of this plant include Botrytis blight, Rhizoctonia canker, Pythium root rot, anthracnose (leaf curl), downy mildew, and tospoviruses.

Diseases of anemone—Anemone mosaic, Botrytis collar rot, crown gall, downy mildew, impatiens necrotic spot, leaf curl, Phytophthora root and crown rot, powdery mildew, Pythium root rot, Rhizoctonia stem and root rot, rust, Sclerotinia blight, Southern blight, and tomato spotted wilt.

Aquilegia caerulea and *A.* × *hybrida* (Ranunculaceae), Columbine

Aquilegia caerulea James and *A.* × *hybrida* Sims are perennials that may be produced as flowering potted plants, cut flowers, or bedding plants for perennial gardens. Bicolors in blue, white, and yellow and solid white and yellow types are available. The flower spurs at the base of the bloom give the blossom a unique character. Seed require chilling before sowing for the best uniformity of germination. As spring-produced bedding plants, columbines will not flower the same season they are seeded unless exposed to low temperatures. Primary arthropod pests include leafminers, whiteflies, and spider mites.

Diseases of columbine—Botrytis blight, foliar nematode, Haplobasidion leaf spot, leaf smut, Phytophthora root rot, a disease caused by an unnamed potyvirus, powdery mildew,

Pythium root rot, Rhizoctonia root and crown rot, root-knot nematode, rust, Sclerotinia blight, Southern blight, Stemphylium leaf spot, and tomato spotted wilt.

Begonia spp. (Begoniaceae), Begonias

The tuberous (nonstop) begonias (*Begonia* Tuberhybrida hybrids) are summer-flowering plants derived from several Andean species. Although traditionally produced vegetatively from tuber divisions, nonstops and other tuberous begonias are now widely grown from seed as bedding plants and specialty items (hanging baskets and finished pots) for spring and summer sales. Night temperatures below 17°C and photoperiods of less than 14 hr are to be avoided to prevent tuber formation and dormancy. Supplemental lighting is necessary during winter months to promote flowering. Spider mites, tarsonemid mites, and thrips are all pests of nonstop begonias. The major disease problems for this group are Botrytis blight, powdery mildew, and impatiens necrotic spot.

The Rieger or hiemalis begonias (*B.* × *hiemalis* Fotsch) are in the Elatior group of the tuberous-rooted begonias. They are hybrids of the winter-flowering, bulbous-rooted *B. socotrana* and the summer-flowering, tuberous-rooted *Begonia* Tuberhybrida cultivars or Andean species. The result is a winter-flowering begonia with large flowers that is produced in greenhouses as a flowering potted plant and is available in shades of red, pink, yellow, orange, and white and in double- and single-flower forms. Plants are produced from leaf or stem-tip cuttings and should be grown at a pH of 5–5.5 in a soilless potting mix. Temperature reduction to 18°C after plants have reached their final size is important for flower initiation and improved quality. Rieger begonias are slightly photoperiodic: long days are supplied for vegetative development followed by short-day treatment to improve the uniformity of flowering and compactness of the finished plant. Plants will not flower well if fertilized excessively and should be shaded during periods of intense light. Plants can tolerate 3,000 footcandles (fc) at 18°C or less, 2,000 fc at 21°C, and only 1,500 fc at 26°C. Broad mites and twospotted spider mites may infest Rieger begonias. Their primary disease problem is powdery mildew, but bacterial blight, Botrytis blight, tospoviruses, and foliar nematodes also commonly cause damage.

The wax begonias (*Begonia* Semperflorens-Cultorum hybrids) are fibrous rooted. They are usually produced from seed in the spring as bedding plants, finished pots, or hanging baskets. Green, red, or bronze foliage contrasts strikingly with pink or white flowers. The major disease problem for this begonia group is Botrytis blight. Fungus gnat larvae and twospotted spider mites may occasionally be problematic.

Diseases of begonia—Alternaria leaf spot, anthracnose, aster yellows, bacterial fasciation, bacterial leaf spot, bacterial soft rot, Botrytis blight, carnation mottle, Cercospora leaf spot, clover yellow mosaic, crown gall, cucumber mosaic, foliar nematode, Fusarium root and crown rot, impatiens necrotic spot, lance nematode, lesion nematode, Myrothecium leaf spot, Phytophthora root rot, potato rot nematode, powdery mildews, Pythium root rot, Rhizoctonia root and crown rot, root-knot nematodes, rust, Sclerotinia blight, southern blight, spiral nematodes, Thielaviopsis root rot, tobacco necrosis, tobacco ringspot, tomato spotted wilt, and Verticillium wilt.

TABLE 1. Common Names of Flowering Potted Plants

Scientific Name	Common Name
Anemone coronaria	Anemone
Aquilegia caerulea,	
A. × *hybrida*	Columbine
Begonia	
× *hiemalis*	Rieger or hiemalis begonia
Semperflorens-Cultorum	
hybrids	Wax begonia
Tuberhybrida hybrids	Nonstop begonia, tuberous begonia
Bougainvillea × *buttiana,*	
B. spectabilis	Bougainvillea
Browallia speciosa	Sapphire flower
Calceolaria Herbeohybrida	
group	Pocketbook plant
Campanula carpatica	Bellflower
Capsicum annuum	
var. *annuum*	Ornamental or Christmas pepper
Catharanthus roseus	Madagascar periwinkle, vinca
Clerodendrum thomsoniae	Bleeding heart vine
Crossandra infundibuliformis	Firecracker-flower
Cyclamen persicum	Florist's cyclamen
Dahlia hybrids	Dahlia
Dicentra spectabilis	Bleeding heart
Euphorbia pulcherrima	Poinsettia
Eustoma grandiflorum	Lisianthus, Texas bluebell
Exacum affine	Persian violet
Fuchsia × *hybrida*	Fuchsia
Gardenia augusta	Common or cape gardenia
Gerbera jamesonii	African daisy
Hatiora gaertneri, H. graeseri,	
H. rosea	Easter cactus
Hibiscus rosa-sinensis	Chinese hibiscus
Hydrangea macrophylla	Florist's hydrangea
Impatiens	
hybrids	New Guinea impatiens
wallerana	Impatiens
Kalanchoe blossfeldiana	Kalanchoe
Lantana camara	Lantana
Mimulus × *hybridus*	Monkey-flower
Pelargonium	
× *domesticum*	Regal or Martha Washington geranium
graveolens	Sweet-scented geranium
× *hortorum*	Florist's geranium
peltatum	Ivy geranium
Pericallis × *hybrida*	Cineraria
Primula	
malacoides	Fairy primrose
obconica	German primrose
Pruhonicensis hybrids	Polyanthus
sinensis	Chinese primrose
vulgaris	English primrose
Ranunculus asiaticus	Ranunculus, Persian buttercup
Saintpaulia ionantha	African violet
Schizanthus × *wisetonensis*	Butterfly flower, poorman's orchid
Schlumbergera	
× *buckleyi*	Christmas cactus
truncata	Thanksgiving cactus
Sinningia speciosa	Gloxinia
Solanum pseudocapsicum	Jerusalem cherry, Christmas cherry
Verbena × *hybrida*	Verbena

Bougainvillea spp. (Nyctaginaceae), Bougainvillea

Woody vines native to tropical South America, bougainvilleas (*Bougainvillea* × *buttiana* Holtt. & Standl. and *B. spectabilis* Willd.) are propagated from cuttings and grown as potted plants or in hanging baskets. Dry conditions are important for flowering, and bloom is fullest when plants are grown under short days. Bougainvillea may be troubled by aphids, twospotted spider mites, and whiteflies.

Diseases of bougainvillea—Alternaria leaf spot, anthracnose, Cercosporidium leaf spot, Fusarium root and crown rot, Nectria dieback, Phytophthora root rot, Pseudomonas leaf spot, Pythium root rot, Rhizoctonia root and crown rot, and tobacco mosaic.

Browallia speciosa (Solanaceae), Sapphire Flower

Browallia speciosa Hook. is a perennial from Colombia and is primarily propagated from seed to produce bedding or hanging-basket plants. It is capable of overwintering outdoors only in frost-free areas. *B. speciosa* produces star-shaped flowers in shades of blue, white, or lavender. It may be attacked by aphids and whiteflies.

Diseases of sapphire flower—Botrytis blight, Fusarium wilt, impatiens necrotic spot, tomato spotted wilt, and Verticillium wilt.

Calceolaria Herbeohybrida group (Scrophulariaceae), Pocketbook Plant

Plants in the *Calceolaria* Herbeohybrida group, sometimes referred to as *C. crenatiflora* Cav., are hybrids grown in the northern United States and northern Europe as low-temperature potted plants for sale during spring holidays. Long days and cool nights are necessary for flowering of older cultivars; newer types are photoinsensitive. Most of the calceolarias grown today are F_1 hybrids. The recommended pH is 5.5–6.0 to avoid foliar chlorosis. These plants require only light fertilization. They are easily overwatered and prone to crown rot if transplanted too deeply. Aphids, thrips, whiteflies, and twospotted spider mites may affect calceolaria. Botrytis blight and tospoviruses are the major disease problems.

Diseases of pocketbook plant—Bacterial leaf spot, Botrytis blight, impatiens necrotic spot, Myrothecium root and crown rot, Phytophthora root rot, Pythium root rot, Sclerotinia blight, tomato spotted wilt, and Verticillium wilt.

Campanula carpatica (Campanulaceae), Bellflower

Campanula carpatica Jacq. is a perennial native to the Carpathians. Its cultivated varieties are lilac blue to white. Cuttings grown outdoors during the summer and allowed to go dormant can be brought into the greenhouse for 10–12 weeks of forcing to yield a spring crop. *C. carpatica* requires long (16-hr) days for flowering. The only disease problem common to campanula is Botrytis crown rot. Whiteflies and aphids are the dominant insect pests.

Diseases of bellflower—Botrytis blight, crown gall, cucumber mosaic, Fusarium root and crown rot, impatiens necrotic spot, a disease caused by an unnamed potyvirus, powdery mildew, Pythium root rot, Southern blight, and tomato spotted wilt.

Capsicum annuum var. annuum (Solanaceae), Ornamental Pepper

Ornamental peppers (*Capsicum annuum* L. var. *annuum*) come from four subgroups of the widely cultivated pepper: the Cerasiforme (cherry peppers), the Conioides (cone pepper), the Fasciculatum (red cherry peppers), and the Longum (chili peppers). All were derived from wild ancestors that ranged from the southern United States to Colombia. They are cultivated from seed. The plants are generally sold from September to Christmas in pots or hanging baskets. Variously shaped fruits range in color from yellow to red. They are quite susceptible to tospoviruses and are attractive to aphids, thrips, and whiteflies.

Diseases of ornamental pepper—Alternaria leaf spot, anthracnose, awl nematode, bacterial leaf spot, bacterial soft rot, bacterial spot, Botrytis blight, burrowing nematode, crown gall, downy mildew, impatiens necrotic spot, lesion nematode, Phytophthora root rot, powdery mildews, Pythium root rot, Rhizoctonia root and crown rot, root-knot nematodes, Sclerotinia blight, sheath nematode, Southern blight, Southern wilt, spiral nematode, sting nematode, stubby-root nematode, tomato spotted wilt, and Verticillium wilt.

Catharanthus roseus (Apocynaceae), Madagascar Periwinkle, Vinca

Catharanthus roseus (L.) G. Don is often referred to as vinca, reminiscent of the old name, *Vinca rosea*. It is also referred to as Madagascar periwinkle after its native territory, which ranges from Madagascar to India. The flower colors include rose pink, mauve, white, and whites with colored eyes. *C. roseus* is cultivated as an annual and sold as a bedding plant or in hanging baskets. Vincas are sensitive to cool conditions, overwatering, and excessively high pH or ammonium levels. Insects other than the western flower thrips are rarely a problem with vinca. Its major disease problems are Rhizoctonia damping-off and Phytophthora stem rot and aerial blight.

Diseases of vinca—Alternaria leaf spot, anthracnose, aster yellows, Botrytis blight, Corynespora leaf spot, crown gall, cucumber mosaic, Fusarium root and crown rot, impatiens necrotic spot, periwinkle chlorotic stunt, periwinkle mosaic, Phyllosticta leaf spot, Phytophthora blight, Phytophthora

crown and root rot, Pseudomonas leaf spot, Pythium root rot, Rhizoctonia root and crown rot, Rhizopus stem rot, root-knot nematode, rust, Southern wilt, and tomato spotted wilt.

Clerodendrum thomsoniae (Verbenaceae), Bleeding Heart Vine

Clerodendrum thomsoniae Balf. is native to west tropical Africa. It may be cultivated from seeds or semihard cuttings. The flowers are borne in cymes and have large, white calyces and crimson or rose magenta corollas. Mites, aphids, whiteflies, and mealybugs may attack clerodendrums. This crop has occasionally been affected by tobacco mosaic virus, impatiens necrotic spot tospovirus, and fungal leaf spots in the United States; in Europe it is known to be susceptible to Botrytis blight, Pythium root rot, and powdery mildew.

Diseases of bleeding heart vine—Botrytis blight, Cercospora leaf spot, Corynespora leaf spot, impatiens necrotic spot, Phyllosticta leaf spot, powdery mildew, Pythium root rot, root-knot nematodes, tobacco mosaic, and tobacco ringspot.

Crossandra infundibuliformis (Acanthaceae), Firecracker-Flower

Crossandra infundibuliformis (L.) Nees is originally from southern India and Sri Lanka. It has bright orange or salmon pink flower spikes and glossy, dark green leaves. Warm conditions are needed for germination (24–29°C) and growth (minimum 18°C at night) of some cultivars. Crossandras are prone to aphids and thrips; few diseases are reported for this crop.

Diseases of firecracker-flower—Alternaria flower spot, foliar nematode, Helminthosporium flower spot, Phytophthora root rot, Pythium root rot, and Rhizoctonia root and stem rot.

Cyclamen persicum (Primulaceae), Cyclamen

Cyclamen persicum Mill. (syns. *C. indicum* and *C. indicum giganteum*) from the eastern Mediterranean region has cordate leaves and twisted corolla lobes in shades of white, pink, violet, or red with a basal purple blotch. Both standard size and miniature forms are available. F_1 hybrids are popular. Cyclamen are seeded in April for Christmas crops. At transplanting, the corms are set above the soil surface to discourage soft rot. The foliage of cyclamen is sensitive to some pesticide treatments containing petroleum distillates. Cyclamen are attractive to aphids, thrips, black vine weevils, and twospotted spider mites. The most common disease problems for cyclamen are Botrytis crown rot, bacterial soft rot, Fusarium wilt, anthracnose, and tospoviruses.

Diseases of cyclamen: Anthracnoses, aster yellows, bacterial leaf spot, bacterial soft rot, Botrytis blight, bulb and stem nematode, cucumber mosaic, Cylindrocarpon root rot, Cylindrocladiella root rot, foliar nematodes, Fusarium wilt, impatiens necrotic spot, lesion nematode, Phoma damping-off, Phyllosticta leaf spot, potato X, powdery mildew, Ramularia stunt and leaf spot, root-knot nematodes, Southern wilt, tobacco mosaic, tomato aspermy, and tomato spotted wilt.

Dahlia hybrids (Asteraceae), Dahlia

Dahlia hybrids are available as numerous cultivars thought to be derived from the crossing of two Mexican species, *D. pinnata* Cav. and *D. coccinea* Cav. Dahlias are propagated from imported tubers or from seed. Aphids, thrips, and mites are common on dahlia. The disease most often seen on tuberous dahlias, tomato spotted wilt, is caused by a tospovirus; seed crops are more likely to be affected by Botrytis blight or white smut.

Diseases of dahlia—Alternaria leaf spot, anthracnoses, Ascochyta leaf spot, aster yellows, bacterial fasciation, bacterial leaf spot, bacterial soft rot, Botrytis blight, Cercospora leaf spot, Choanephora blossom blight, Colletotrichum root and stem rot, crown gall, cucumber mosaic, dahlia mosaic, foliar nematodes, Fusarium root and crown rot, Fusarium wilt, impatiens necrotic spot, Itersonilia flower blight, lesion nematode, needle nematode, Phyllosticta leaf spot, potato rot nematode, potato Y, powdery mildews, Pythium root rots, Rhizoctonia root and crown rot, root-knot nematodes, Sclerotinia blight, Southern blight, Southern wilt, stubby-root nematode, tobacco ringspot, tobacco streak, tomato spotted wilt, Verticillium wilt, and white smut.

Dicentra spectabilis (Fumariaceae), Bleeding Heart

Dicentra spectabilis (L.) Lem. is a perennial occasionally grown as a potted plant. Petals are white or rosy red with white. Aphids and whiteflies are the primary insect problems in greenhouse culture. Botrytis blight is the only common disease.

Diseases of bleeding heart—Anthracnose, Botrytis blight, Cercospora leaf spot, Stemphylium leaf spot, and Verticillium wilt.

Euphorbia pulcherrima (Euphorbiaceae), Poinsettia

Euphorbia pulcherrima Willd. ex Klotzsch is a winter-flowering, woody plant grown primarily for Christmas sales. Cultivars are available in bract colors of white, pink, yellow, and marbled mixes as well as the original vermilion and additional red hues. Pot plants are produced from cuttings. The major insect problems for poinsettia are the silver leaf whitefly and the greenhouse whitefly; they are also troubled by fungus gnat larvae. The primary diseases are root and stem rots and Botrytis blight. A new powdery mildew disease has recently appeared in North American greenhouses.

Diseases of poinsettia—Alternaria leaf spot, anthracnose, Ascochyta leaf spot, bacterial cankers, bacterial soft rots, black root rot, Botrytis blight, burrowing nematode, Cercospora leaf spot, Choanephora stem rot, Corynespora leaf and bract spot, crinkle mosaic, crown gall, enation leaf curl, Fusarium root and crown rots, greasy canker, Helminthosporium leaf spot, Lasiodiplodia dieback, Nectriella stem gall, Phomopsis stem canker, Phyllosticta leaf spot, Phytophthora root rot, poinsettia cryptic, poinsettia mosaic, powdery mildews, Pythium root rot, Rhizoctonia root and crown rot, Rhizopus root and stem rot, root-knot nematode, rust, scab, Syspatospora leaf and stem rot, Verticillium wilt, and Xanthomonas leaf spot.

Eustoma grandiflorum (Gentianaceae), Lisianthus or Texas Bluebell

Eustoma grandiflorum (Raf.) Shinn. is the annual-biennial commonly referred to as lisianthus. It is used as a potted plant, cut flower, or bedding plant. Native to the United States from Colorado and Nebraska south to Texas and northern Mexico, it has been developed as a potted plant in the Orient. Its flowers are blue, lilac, rose, pink, or white. Mites, whiteflies, leafminers, thrips, and aphids may infest lisianthus. The most common disease problems for this crop during greenhouse production are Botrytis blight, Pythium root rot, and Fusarium wilt.

Diseases of lisianthus—Bean yellow mosaic, Botrytis blight, Cercospora leaf spot, cucumber mosaic, downy mildew, Fusarium stem rot, Fusarium wilt, impatiens necrotic spot, lisianthus necrosis, Phyllosticta leaf spot, Pythium root rot, Sclerophoma stem blight, and tobacco mosaic.

Exacum affine (Gentianaceae), Persian Violet

Exacum affine Balf. f. is from the Indian Ocean island of Socotra. It is an annual or biennial and flowers in shades of violet blue to white. Plants are grown from seed. Exacum are prone to copper deficiency. Broad mites, thrips, aphids, and whiteflies may be damaging. Exacum grown at low light levels (during the winter) may be especially susceptible to Botrytis cankers; the plant is also very susceptible to the impatiens necrotic spot tospovirus.

Diseases of Persian violet—Botrytis blight, impatiens necrotic spot, Nectria canker, Phytophthora blight, and Pythium root rot.

Fuchsia × *hybrida* (Onagraceae), Fuchsia

Fuchsia × *hybrida* Hort. ex Vilm. is thought to be derived from *F. fulgens* and *F. magellanica*, native to Mexico and South America, respectively. The plant is a shrub with only slight frost tolerance. It is generally propagated from soft tip cuttings but is sometimes grown from seed. It is a cool-weather plant used widely for hanging baskets. Fuchsias flower under long-day conditions. They are commonly troubled by aphids, whiteflies, thrips, or cyclamen mites. The most common disease problems on fuchsia are Botrytis cankers, Phytophthora blight, rust, and root rots.

Diseases of fuchsia—Botrytis blight, Cercospora leaf spot, crown gall, impatiens necrotic spot, Phytophthora crown rot, Pythium root rots, Rhizoctonia root and crown rot, root-knot nematode, rust, Southern wilt, Thielaviopsis root rot, and Verticillium wilt.

Gardenia (Rubiaceae), Common or Cape Gardenia

Gardenia augusta (L.) Merrill (syn. *G. jasminoides* J. Ellis) is an evergreen shrub from China with highly scented, creamy white flowers. It is produced in greenhouses primarily in the southern United States. The gardenia's main disease problems are Phytophthora root rot, Rhizoctonia stem cankers, and fungal leaf spots. Aphids (particularly the melon aphid) are the most serious insect pests; western flower thrips, twospotted spider mites, mealybugs, and whiteflies are occasionally problems. The plant is also susceptible to tospoviruses.

Diseases of gardenia—Alternaria leaf spot, anthracnose, Botrytis blight, burrowing nematode, Cercospora leaf spot, Cylindrocladium root and crown rot, Diplodia tip dieback, Erythricium limb blight, foliar nematode, Fusarium bud rot, Fusarium root rot, impatiens necrotic spot, Leptothyrium tip necrosis, Myrothecium leaf spot, Phoma canker, Phomopsis canker, Phyllosticta leaf spot, Phytophthora root and crown rot, powdery mildew, Pythium root rot, Rhizoctonia root and stem rot, root-knot nematode, Sphaeropsis leaf spot, spiral nematodes, tomato spotted wilt, and Xanthomonas leaf spot.

Gerbera jamesonii (Asteraceae), African Daisy

Gerbera jamesonii H. Bolus ex Adlam is a brightly colored daisy from Africa. The cultivated forms are available with ray flowers in deep shades of yellow, salmon, pink, and red. African daisies are now propagated from tissue culture as well as from seed. They are prone to boron, magnesium, and iron deficiencies when grown in soilless mixes. They do best at high light intensities. The major insect problems for African daisy are aphids, leafminers, thrips, and whiteflies; tarsonemid mites may also be damaging. The most common diseases are a bacterial leaf spot, Phytophthora and Pythium root rots, and powdery mildew.

Diseases of gerbera—Alternaria leaf spot, Ascochyta leaf spot, aster yellows, Botrytis blight, Cercospora leaf spot, cucumber mosaic, downy mildew, foliar nematode, Fusarium root and crown rot, Fusarium wilt, impatiens necrotic spot, Phyllosticta leaf spot, Phytophthora root and crown rot, powdery mildew, Pseudomonas leaf spot, Pythium root rot, Rhizoctonia root and crown rot, Rhizopus blight, root-knot nematodes, Sclerotinia blight, Septoria leaf spot, Southern blight, Southern wilt, sting nematode, stubby-root nematode, Thielaviopsis root rot, tobacco mosaic, tobacco rattle, tomato spotted wilt, Verticillium wilt, and white rust.

Hibiscus rosa-sinensis (Malvaceae), Chinese Hibiscus

Hibiscus rosa-sinensis L. is a shrub probably native to tropical Asia. Cultivars are available with flowers of red, pink, yellow, and orange and with variegated foliage. Hibiscus is limited to indoor (and summer patio) use in areas with frost. Propagation of desirable hybrids is by cuttings. Flowering of hibiscus is not photoperiodic, but plants will flower best with high light levels and long days. Aphids, whiteflies, twospotted spider mites, leafminers, and thrips may affect hibiscus. The most common disease problems for hibiscus are bacterial and fungal leaf spots.

Diseases of Chinese hibiscus—Alternaria leaf spot, anthracnoses, Botryosphaeria canker, Botryosphaeria leaf spot, Botrytis blight, burrowing nematode, Cercospora leaf spot, Choanephora blossom blight, Coniothyrium dieback, Curvularia leaf spot, foliar nematode, Fusarium wilt, hibiscus chlorotic ringspot, hibiscus latent ringspot, impatiens necrotic spot, Leptosphaeria leaf spot, Macrophoma leaf spot, Phomopsis leaf spot and dieback, Phyllosticta leaf spot, Physalospora leaf spot, Phytophthora root and crown rot, Pseudocercospora leaf spot, Pseudomonas leaf spot, Pythium root rot, a disease caused by an unnamed rhabdovirus, Rhizoctonia root and stem rot, root-knot nematodes, Sclerotinia crown rot, Southern blight, Sphaeropsis witches'-broom, Volutella canker, and Xanthomonas leaf spot.

Hydrangea macrophylla (Hydrangeaceae), Florist's Hydrangea

Hydrangea macrophylla (Thunb.) Ser., the florist's hydrangea, is native to Japan. Cultivars used for greenhouse forcing are generally not hardy in areas colder than USDA hardiness zone 7. The inflorescence is a globular cyme with large, showy, petaloid sepals on the staminate flower. Plants are usually propagated from cuttings during May and June and given a cold-storage treatment for 6–8 weeks before forcing in the greenhouse under cool night conditions. Through cultivar selection and nutrient regulation, white-, pink-, and blue-sepaled forms may be produced. Hydrangeas commonly show iron deficiency. Plants should be carefully watered to avoid wilting during production. Aphids, whiteflies, leafminers, and spider mites often affect hydrangea. The most common disease problems encountered in hydrangea production are Botrytis blight, powdery mildew, and hydrangea ringspot.

Diseases of hydrangea—Alfalfa mosaic, Alternaria leaf spot, anthracnoses, black root rot, Botrytis blight, burrowing nematode, Cercospora leaf spot, Corynespora leaf spot, crown gall, cucumber mosaic, foliar nematode, Helminthosporium leaf spot, hydrangea latent, hydrangea mosaic, hydrangea ringspot, hydrangea virescence, impatiens necrotic spot, Nectriella stem gall, Phoma leaf spot, Phomopsis canker, Phyllosticta leaf spot, Phytophthora root rot, powdery mildews, Pseudomonas leaf spot, Pythium root rot, Rhizoctonia root and crown rot, root-knot nematodes, Sclerotinia blight, Southern blight, Southern wilt, tobacco necrosis, tobacco ringspot, and tomato spotted wilt.

Impatiens wallerana and Impatiens hybrids (Balsaminaceae), Impatiens, New Guinea Impatiens

Impatiens wallerana Hook. f., the common, seed-propagated bedding plant, comes from Tanzania and Mozambique, where it is a succulent, perennial herb. The lower sepal is elongated and has a curved spur. There are many color forms in pink, white, violet, and red. Single-flowered forms are produced from seed; rosebud types are generally grown from cuttings. New Guinea impatiens, grown from cuttings (or, rarely, seed), are hybrids of *Impatiens* spp. from New Guinea. They also have a succulent stem, but their leaves are darker green and more elongated and their flowers are larger than those of *I. wallerana*. Both plants are widely grown as bedding plants and hanging basket items. *I. wallerana* does better in shade, and New Guinea impatiens will withstand full sun in areas where summer temperatures are not extremely high. The most important arthropod pests on impatiens are twospotted spider mites, tarsonemid mites, and thrips. Both types of impatiens are often injured by *Rhizoctonia solani*, bacterial leaf spots, and tospoviruses, particularly impatiens necrotic spot virus; New Guinea Impatiens are also often affected by Pythium root rot.

Diseases of impatiens—Bacterial fasciation, Botrytis blight, cucumber mosaic, downy mildew, Fusarium root and crown rot, Helenium S, impatiens necrotic spot, Myrothecium leaf spot, Phyllosticta leaf spot, Phytophthora root rot, a disease caused by an unnamed potexvirus, powdery mildew, Pseudomonas leaf spot, Pythium root rot, Rhizoctonia root and stem rot, root-knot nematode, Southern blight, tobacco streak, tomato spotted wilt, and Verticillium wilt.

Kalanchoe blossfeldiana (Crassulaceae), Kalanchoe

Kalanchoe blossfeldiana Poelln. is a succulent perennial herb from Madagascar with a corymbose inflorescence. The plant naturally flowers during the winter in intense shades of yellow, orange, pink, or red; but in greenhouse culture, its flowering is induced all year long by providing short days (14-hr nights). Plants are most often produced from cuttings but may also be produced from seed. The foliage of kalanchoes is sensitive to some pesticide treatments containing petroleum distillates. Aphids may be a problem. The diseases most often encountered in kalanchoe production are bacterial stem rot and powdery mildew.

Diseases of kalanchoe—Bacterial fasciation, bacterial soft rots, Botrytis blight, Cercospora leaf spot, crown gall, Cylindrocladium root rot, Fusarium crown rot, impatiens necrotic spot, kalanchoe mosaic, kalanchoe top-spotting, KV-1 and KV-2 latent, Phytophthora crown rot, a disease caused by an unnamed potyvirus, powdery mildews, Pythium root rot, Rhizoctonia root and crown rot, Stemphylium leaf spot, tomato bushy stunt, and tomato spotted wilt.

Lantana camara (Verbenaceae), Lantana

Lantana camara L. is a hairy shrub from the tropical Americas naturalized in the southern United States and Hawaii. It is a tender perennial produced primarily from softwood cuttings or from seed. Vegetatively propagated lantanas come in a diversity of flower colors, including cream, pink, orange, yellow, and purple. The color range is limited in seed-propagated varieties. This plant is highly attractive to whiteflies. The most common disease problem for lantana in the greenhouse is Rhizoctonia stem rot.

Diseases of lantana—Botrytis blight, crown gall, foliar nematode, impatiens necrotic spot, lesion nematode, Mycovellosiella leaf spot, Phytophthora flower blight, Rhizoctonia stem rot, root-knot nematodes, rust, Southern wilt, tomato spotted wilt, and Xanthomonas leaf spot.

Mimulus × hybridus (Scrophulariaceae), Mimulus, Monkey-Flower

Mimulus × hybridus hort. ex Siebert & Voss is a cross between *M. luteus* and *M. guttatus* whose large, two-lipped, spotted flowers suit it to pot-plant culture. Its parent plants are native to North and South America. Mimulus is produced

from seed sown from January to April. It requires long days for flowering and grows best outdoors in cool, shaded areas. Its most common insect problems are aphids and whiteflies; plants are also susceptible to twospotted spider mites. Mimulus is susceptible to impatiens necrotic spot, Botrytis blight, and Pythium root rot.

Diseases of mimulus—Aster yellows, Botrytis blight, crown gall, downy mildew, impatiens necrotic spot, powdery mildews, and Pythium root rot.

Pelargonium spp. (Geraniaceae), Florist's Geranium, Ivy Geranium, Regal Geranium

Pelargonium × hortorum L. H. Bailey (the florist's geranium, sometimes called *P. zonale*), *P. peltatum* (L.) L'Hér. (the ivy geranium), and *P. × domesticum* L. H. Bailey (the Martha Washington or regal geranium) are very popular greenhouse potted plants that flower in hues of red, orange, pink, white, and violet. *P. graveolens* L'Hér. (the sweet-scented geranium) is grown as an ornamental specialty crop and is also cultivated as a perfume ingredient. *P. × hortorum* is produced both from cuttings and from F₁ hybrid seed and is also used as a bedding plant. *P. peltatum* is primarily produced from cuttings and is used in hanging baskets. *P. × domesticum* is a potted plant produced from cuttings. *P. peltatum* is commonly affected by a physiological edema and is attractive to thrips; it is also commonly a host of the twospotted spider mite. *P. × domesticum* is often attacked by whiteflies. The most common disease problems of *Pelargonium* spp. are Botrytis blight, Pythium root rot, and Pelargonium flower break. Bacterial blight (on *P. hortorum* and *P. peltatum*) and rust (on *P. hortorum*) occur less frequently but cause major crop losses.

Diseases of geranium—Alfalfa mosaic, Alternaria leaf spot, anthracnose, arabis mosaic, artichoke Italian latent, bacterial blight, bacterial fasciation, bacterial soft rot, beet curly top, Bipolaris leaf and petal spot, Botrytis blight, Cercospora leaf spot, crown gall, cucumber mosaic, Cylindrocladium leaf spot and canker, dagger nematodes, foliar nematodes, Fusarium root rot, Fusarium stem rot, impatiens necrotic spot, lesion nematode, Pelargonium flower break, Pelargonium leaf curl (caused by a strain of tomato bushy stunt tombusvirus), Pelargonium ring pattern, Pelargonium veinclearing, Pelargonium yellow net vein, Pelargonium zonate spot, Phyllosticta leaf spot, Phytophthora root rot, Pseudomonas leaf spot, Pythium root rot, Rhizoctonia root and crown rot, root-knot nematodes, rust, Sclerotinia blight, Southern blight, Southern wilt, stunt nematode, Thielaviopsis root rot, tobacco mosaic, tobacco necrosis, tobacco rattle, tobacco ringspot, tomato black ring, tomato ringspot, tomato spotted wilt, and Verticillium wilt.

Pericallis × hybrida (Asteraceae), Cineraria

Pericallis × hybrida R. Nordenstam (syn. *Senecio × hybridus* (Willd.) Regel), the florist's cineraria, is a perennial usually cultivated as an annual produced from seeds. Its flowers come in intense solid colors as well as with white eyes. The crop is grown at 7–20°C, and flower initiation is achieved after a month of growing temperatures of 7–13°C. Cinerarias are plagued by thrips, whiteflies, twospotted spider mites, and especially aphids. Their greatest disease problem is impatiens necrotic spot.

Diseases of cineraria—Alternaria leaf spot, aster yellows, Botrytis blight, crown gall, downy mildew, foliar nematode, impatiens necrotic spot, Phytophthora root and crown rot, powdery mildews, Pythium root rot, root-knot nematode, rusts, Southern wilt, tomato spotted wilt, Verticillium wilt, and white rust.

Primula spp. (Primulaceae), Primrose

The primulas commonly grown as ornamentals include *Primula vulgaris* Huds. (syn. *P. acaulis*), English primrose; *Primula* Pruhonicensis hybrids, polyanthus; *P. malacoides* Franch., fairy primrose; *P. obconica* Hance, German primrose; and *P. sinensis* Sabine ex Lindl., Chinese primrose.

Most greenhouse-grown primulas originated in China, except for *P. vulgaris,* which is native to southeastern Europe, the Caucasus region, and Iran. Polyanthus are hybrids of many different species. The various primroses (*Primula* spp.) are all perennial herbs produced from seed as pot crops; *P. vulgaris* and *P. malacoides* are also used as bedding plants. *P. sinensis, P. acaulis,* and polyanthus have brilliant, clear colors (bicolors and F₁ hybrids are available for *P. acaulis*), and *P. obconica* and *P. malacoides* offer pastels. *P. malacoides, P. sinensis,* and *P. obconica* are all sensitive to high levels of soluble salts. Aphids and twospotted spider mites are common on primulas. The most common disease problem for this plant group is Botrytis crown rot; they are also prone to Ramularia and Pseudomonas leaf spots.

Diseases of primrose—Alfalfa mosaic, anthracnose, aster yellows, bacterial fasciation, Botrytis blight, bulb and stem nematode, crown gall, cucumber mosaic, downy mildew, foliar nematodes, impatiens necrotic spot, lesion nematode, Phyllosticta leaf spot, Phytophthora crown and root rot, powdery mildew, Primula mosaic, Primula mottle, Pseudomonas leaf spot, Pythium root rot, Ramularia leaf spot, Rhizoctonia root and crown rot, root-knot nematodes, rusts, tobacco necrosis, tomato bushy stunt, and tomato spotted wilt.

Ranunculus asiaticus (Ranunculaceae), Ranunculus, Persian Buttercup

Ranunculus asiaticus L. is a native of southeastern Europe and southwestern Asia. Plant-breeding efforts have resulted in the development of seed-propagated genetic dwarfs that perform well as potted plants. Ranunculus may also be propagated from tubers. A 30- to 40-day cold treatment allows production of a flowering crop from November to early June. Seed germination must be carried out under evenly moist, cool (10–15°C) conditions. Typically, seed is sown on September 1 in cool climates and on October 1 in warm areas for a spring crop. After they are transplanted, ranunculus are grown under conditions of short days and low temperatures (12–15°C days and 5°C nights). Once flower buds are visible, temperatures are raised to 15–20°C during the day and 7–10°C at night. Aphids, leafminers, and spider mites commonly attack ranunculus. The major disease problems in ranunculus production are powdery mildew, Fusarium wilt, root rots, and impatiens necrotic spot.

Diseases of ranunculus—Anthracnose, aster yellows, Botrytis blight, bulb and stem nematodes, crown gall, downy mildews, foliar nematodes, Fusarium wilt, impatiens necrotic spot, lesion nematode, potato Y, powdery mildews, Pythium root rot, disease caused by Ranunculus potyvirus, Rhizoctonia stem and root rot, root-knot nematode, rust, and tomato spotted wilt.

Saintpaulia ionantha (Gesneriaceae), African Violet

Saintpaulia ionantha Wendl. is a perennial, hairy herb that originated in coastal Tanzania. It has been developed into many cultivars with flowers in hues of purple, pink, violet, and white as well as bicolors. These plants are grown primarily from leaf cuttings year round in greenhouses under warm conditions in diffused light (1,500 fc). African violets require only light fertilization and are sensitive to high levels of soluble salts. Cyclamen mites, mealybugs, and thrips may infest African violets. The most common disease problems are Botrytis blight; foliar nematodes; powdery mildew; and Phytophthora, Pythium, and Rhizoctonia root and crown rots.

Diseases of African violet—Alternaria leaf spot, bacterial soft rot, Botrytis blight, Corynespora leaf spot, crown gall, Cylindrocarpon root and crown rot, foliar nematodes, Fusarium root rot, impatiens necrotic spot, lance nematode, lesion nematode, Phytophthora root and crown rot, powdery mildews, Pythium root and crown rot, Rhizoctonia root and crown rot, ring nematode, root-knot nematode, spiral nematode, tobacco mosaic, and tomato spotted wilt.

Schizanthus × *wisetonensis* (Solanaceae), Butterfly Flower, Poorman's Orchid

Schizanthus × *wisetonensis* hort., native to Chile, is grown under cool greenhouse conditions from seed. The flowers range from white to blue or pink to carmine brown, usually with contrasting yellow centers. Whiteflies, western flower thrips, and aphids are problems. The most common disease problems are *Pythium* and *Botrytis* infections.

Diseases of butterfly flower—Anthracnose, Ascochyta leaf spot, aster yellows, bacterial fasciation, Botrytis blight, crown gall, foliar nematode, impatiens necrotic spot, Phytophthora root rot, root-knot nematode, and tomato spotted wilt.

Schlumbergera and *Hatiora* spp. (Cactaceae), Holiday Cactus

The holiday cacti are epiphytes native to Brazil and have jointed stems made up of a series of flattened cladophylls. The majority of the holiday cactus clones grown commercially in North America are *Schlumbergera truncata* (Haw.) Moran (formerly *Zygocactus truncatus* (Haw.) K. Schum.), the Thanksgiving (or crab) cactus. These have dentate cladophylls and a strongly zygomorphic flower and bloom naturally in mid- to late November. *S. truncata* is often referred to as Christmas cactus rather than Thanksgiving cactus.

Christmas cacti, *S.* × *buckleyi* (T. Moore) Tjaden. (formerly *S. bridgesii* (Lem.) Loefgr.), are interspecific hybrids that bloom roughly 1 month later than *S. truncata.* They have pendulous blossoms that are only partially zygomorphic, and the cladophylls are crenate.

Easter cacti are species and interspecific hybrids of *Hatiora*: *H. gaertneri* (Regel) Barthlott (formerly *Rhipsalidopsis gaertneri* (Regel) Moran), *H. rosea* (Lagerheim) Barthlott (formerly *R. rosea* (Lagerheim) Britton & Rose), and *H.* × *graeseri* (Werdermann) Barthlott (formerly *R.* × *graeseri* (Werdermann) Moran). Easter cactus has an actinomorphic flower and cladophyll margins with varying degrees of crenation.

Propagation of holiday cactus is by cuttings. The flower buds of *Schlumbergera* spp. are initiated by either short days (less than 12 hr) or low temperatures (15–20°C). Fungus gnat larvae may be a problem during propagation. The main disease problems for holiday cactus are Pythium root rot and infection of the cladophylls by *Fusarium* and *Bipolaris* spp.

Diseases of holiday cactus—Anthracnoses, bacterial soft rot, Bipolaris leaf spot, Cercospora leaf spot, Dichotomophthora leaf spot, Fusarium stem rot, impatiens necrotic spot, lesion nematode, Phomopsis leaf spot, Phytophthora root rot, Pythium root rot, and tomato spotted wilt.

Sinningia speciosa (Gesneriaceae), Gloxinia

Sinningia speciosa (Lodd.) Hiern is a tender, perennial herb with hairy leaves from Brazil. Cultivars offer large, bell-shaped flowers in violet, rose red, white, or yellow, some marked with spots. Although gloxinias may be produced from tubers, they are usually grown from seed under warm conditions (18°C nights and 24°C days) and slightly higher light intensity than African violets (2,000–2,500 fc). Cyclamen mites, aphids, whiteflies, leafminers, and thrips may trouble gloxinias. Their most common disease problems are Phytophthora crown rot and tospoviruses.

Diseases of gloxinia—Anthracnose, aster yellows, Botrytis blight, Cladosporium leaf spot, foliar nematode, impatiens necrotic spot, Myrothecium leaf spot and crown rot, Phytophthora root and crown rot, Plenodomus leaf spot, Pythium root rot, Rhizoctonia root and crown rot, root-knot nematodes, Sclerotinia blight, Southern blight, tobacco mosaic, tobacco ringspot, and tomato spotted wilt.

Solanum pseudocapsicum (Solanaceae), Jerusalem Cherry, Christmas Cherry

Solanum pseudocapsicum L. is an Old World plant now naturalized in the tropics and subtropics, including USDA zone 9 in the United States. A pot-plant crop is seeded in February, transplanted and grown outdoors during the summer, and returned to the greenhouse for cool (10–13°C) finishing for Christmas sales. Aphid and thrips infestations may occur on Jerusalem cherry. Tospoviruses are the most common disease threat for this crop.

Diseases of Jerusalem cherry—Alternaria leaf spot, anthracnose, Botrytis blight, cucumber mosaic, cyst nematode, impatiens necrotic spot, Phyllosticta leaf spot, Phytophthora

root and crown rot, powdery mildew, Pythium root rot, Rhizoctonia root and crown rot, root-knot nematodes, Southern blight, Stemphylium leaf spot, tobacco mosaic, tomato spotted wilt, and Verticillium wilt.

Verbena × *hybrida* (Verbenaceae), Verbena

Verbena × *hybrida* Groenl. & Ruempl. (syn. *V.* × *hortensis* Vilm.) is a perennial used as a potted plant or annual bedding plant. Both creeping and upright forms are available. The flower spikes are flattened into heads; the flowers are pink, red, white, yellow, blue, purple, or variegated. Seed is sown during February or March for spring sales. The main insect problems on verbena are aphids and whiteflies; the primary disease is powdery mildew.

Diseases of verbena—Alternaria leaf spot, anthracnose, bidens mottle, Botrytis blight, Cercospora leaf spot, impatiens necrotic spot, a disease caused by an unnamed potyvirus, powdery mildew, Rhizoctonia root and crown rot, Thielaviopsis root rot, and tomato spotted wilt.

Selected References

Agrios, G. N. 1988. Plant Pathology. 3rd ed. Academic Press, New York.

Alfieri, S. A., Jr., Langdon, K. R., Wehlburg, C., and Kimbrough, J. W. 1984. Index of plant diseases in Florida. Fla. Dep. Agric. Consumer Serv. Div. Plant Indus. Bull. 11.

Bailey, L. H., Bailey, E. Z., and the Staff of the L. H. Bailey Hortorium. 1976. Hortus Third. Macmillan, New York.

Ball, V., ed. 1991. Ball Red Book. 15th ed. George J. Ball, West Chicago, IL.

Bradbury, J. F. 1986. Guide to Plant Pathogenic Bacteria. C. A. B. International, Slough, England.

Chase, A. R. 1987. Compendium of Ornamental Foliage Plant Diseases. American Phytopathological Society, St. Paul, MN.

Ecke, P. E., Jr., Matkin, O. A., and Hartley, D. E. 1990. The Poinsettia Manual. 3rd ed. Paul Ecke Poinsettias, Encinitas, CA.

Farr, D. F., Bills, G. F., Chamuris, G. P., and Rossman, A. Y. 1989. Fungi on Plants and Plant Products in the United States. American Phytopathological Society, St. Paul, MN.

Forsberg, J. L. 1975. Diseases of Ornamental Plants. 2nd ed. University of Illinois Press, Urbana.

Goodey, T., Franklin, M. T., and Hooper, D. J. 1965. The Nematode Parasites of Plants Catalogued under Their Hosts. 3rd ed. C. A. B. International, Slough, England.

Griffiths, M. 1994. Index of Garden Plants. Timber Press, Portland, OR.

Horst, R. K. 1990. Westcott's Plant Disease Handbook. 5th ed. Van Nostrand Reinhold, New York.

Pirone, P. P. 1978. Diseases and Pests of Ornamental Plants. 5th ed. John Wiley & Sons, New York.

Strider, D. L., ed. 1985. Diseases of Floral Crops, vols. 1 and 2. Praeger Scientific, New York.

von Hentig, W. U., Röber, R., Wohanka, W., and Rohde, J. 1991. Guida alla Coltivazione delle Piante Ornamentali. G. Rampini, ed. Pentagono Editrice Sas, Milan, Italy.

White, J. W., ed. 1993. Geraniums IV. Ball Publishing, Geneva, IL.

Part I. Infectious Diseases

Diseases Caused by Fungi

Fungi are heterotrophic organisms that generally have highly branched, tubular bodies and produce spores for the dual purposes of reproduction and dissemination. They exude enzymes that break down an external substrate so that nutrients can be absorbed. Plants parasitized by pathogenic fungi exhibit a wide range of symptoms, including leaf spots, blossom blights, stem cankers, galls, root and crown rots, and vascular wilts. The pathogens that cause rust and powdery mildews may themselves be observed on infected plant tissue.

Fungal pathogens are generally introduced via infested plant material (e.g., plants and seeds) or soil particles blown or tracked into the greenhouse. Occasionally, spores may be blown directly into the greenhouse from outside, or insects entering the greenhouse may vector fungi. Inoculum of a fungus may be retained in the greenhouse in the form of resting structures such as oospores, chlamydospores, and sclerotia. These structures allow a fungus to remain dormant in the soil when a suitable substrate is lacking or environmental conditions are unfavorable.

The following sections cover some of the more important fungal diseases of flowering potted plants. Other diseases that have been reported but not studied in depth, such as Phyllosticta leaf spots and other miscellaneous diseases, are listed in tables.

Selected References

Alexopoulos, C. J., and Mims, C. W. 1979. Introductory Mycology. John Wiley & Sons, New York.

Ellis, M. B. 1971. Dematiaceous Hyphomycetes. Commonwealth Mycological Institute and Commonwealth Agricultural Bureaux, Kew, England.

Ellis, M. B. 1976. More Dematiaceous Hyphomycetes. Commonwealth Mycological Institute and Commonwealth Agricultural Bureaux, Kew, England.

Ellis, M. B., and Ellis, J. P. 1985. Microfungi on Land Plants. Macmillan, New York.

Farr, D. F., Bills, G. F., Chamuris, G. P., and Rossman, A. Y. 1989. Fungi on Plants and Plant Products in the United States. American Phytopathological Society, St. Paul, MN.

Hanlin, R. T. 1990. Illustrated Genera of Ascomycetes. American Phytopathological Society, St. Paul, MN.

Sutton, B. C. 1980. The Coelomycetes. Commonwealth Mycological Institute and Commonwealth Agricultural Bureaux, Kew, England.

von Arx, J. A. 1987. Plant Pathogenic Fungi. J. Cramer, Berlin.

Leaf Spots and Blights

Alternaria Leaf Spot of Geranium

Alternaria leaf spot on *Pelargonium* × *hortorum* L. H. Bailey and *P. peltatum* (L.) L'Hér. has been reported from Florida, California, and Europe. It has also been observed on *P.* × *domesticum* L. H. Bailey and *P. graveolens* L'Hér. in the United States. The disease occurs most frequently under production conditions stressful to the host.

Symptoms

Small (1–2 mm), blisterlike, water-soaked spots first appear on the undersides of older, lower geranium leaves. As the spots mature, the centers become sunken, brown, and 2–3 mm in diameter and may show diffuse, yellow halos (Plate 1). Eventually, spotting is apparent on the upper leaf surfaces as well. If conditions remain suitable for disease development, spots enlarge to irregularly shaped areas 6–10 mm in diameter. In some cases, the lesions show concentric rings. Neighboring lesions may coalesce. Chlorosis of infected leaves and leaf abscission may occur. Dark spores are produced on the necrotic spots, particularly on fallen leaves. The absence of associated wilting and the large size range of spots (1–10 mm) help to distinguish this disease from bacterial blight caused by *Xanthomonas campestris* pv. *pelargonii*.

Causal Organism

Alternaria alternata (Fr.:Fr.) Keissl. (syn. *A. tenuis* Nees) has simple or branched, brown conidiophores (up to 50 × 3–6 μm) that are grouped or single. The muriform, light golden brown conidia (20–63 × 9–18 μm) are produced in long chains that may be branched. They are smooth or rough walled and have very short, pale beaks. The colonies are effuse and range in color from gray to black or olivaceous black. *A. alternata* is a widely distributed saprophyte that occasionally causes diseases of plants.

Host Range and Epidemiology

A. alternata has several potted-plant hosts, including *Dahlia* hybrids, *Gerbera jamesonii* H. Bolus ex Adlam (African daisy), *Catharanthus roseus* (L.) G. Don (vinca) (Table 2 and Plate 2), and *Hibiscus rosa-sinensis* L. (hibiscus). Many citations of Alternaria leaf spot occurrence identify the pathogen only as *Alternaria* sp. (Table 3). The wide host range of *A. alternata* suggests that this species may be responsible for some of the diseases attributed to *Alternaria* sp.

In Florida, Alternaria leaf spot on geraniums usually occurs in months during which the temperature is above the optimum

TABLE 2. Relative Susceptibility of Vinca (*Catharanthus roseus*) Cultivars to Alternaria Leaf Spot[a]

Low	Medium	High
Tropicana Rose	Tropicana Blush	Tropicana Pink
Tropicana Bright Eye	Parasol	Cooler Grape
	Little Blanche	Cooler Peppermint
		Cooler Blush

[a] Data from Chase, 1994.

9

TABLE 3. *Alternaria* spp. that Cause Leaf or Flower Spots on Flowering Potted Plants

Host	*Alternaria* sp.
Begonia sp.	*Alternaria* sp.
Bougainvillea sp.	*Alternaria* sp.
Capsicum annuum (ornamental pepper)	*A. solani*
Catharanthus roseus (vinca)	*A. alternata*
Crossandra infundibuliformis	*Alternaria* sp.
Dahlia hybrids	*A. alternata*
Euphorbia pulcherrima (poinsettia)	*A. euphorbiicola*
Gardenia augusta	*Alternaria* sp.
Gerbera jamesonii (African daisy)	*A. alternata,* *A. dauci,* *A. gerberae*
Hibiscus rosa-sinensis	*A. alternata*
Hydrangea macrophylla	*Alternaria* sp.
Pelargonium spp. (geranium)	*A. alternata*
Pericallis × hybrida (cineraria)	*A. cinerariae*
Saintpaulia ionantha (African violet)	*Alternaria* sp.
Schizanthus × wisetonensis (butterfly flower)	*Alternaria* sp.
Schlumbergera truncata (Thanksgiving cactus)	*Alternaria* sp.
Solanum pseudocapsicum (Jerusalem cherry)	*A. solani*
Verbena × hybrida	*Alternaria* sp.

for geranium growth. However, in California, the disease is most common at temperatures that are lower than optimal during foggy, rainy weather. The disease may also develop in closed boxes during shipment.

Management

Infected leaves should be gathered to reduce inoculum, and temperature stress and extended periods of leaf wetness should be avoided. Wide temperature fluctuations during shipping and retention of plants in boxes for extended periods should also be avoided. Chlorothalonil and iprodione are registered in the United States for control of *A. alternata* on geranium. Fungicide labels give additional crop-use specifications.

Selected References

Chase, A. R. 1994. Resistance of vinca cultivars to Alternaria leaf spot, 1993. Biol. Cult. Tests Control Plant Dis. 9:166.

Ellis, M. B. 1971. Dematiaceous Hyphomycetes. Commonwealth Mycological Institute and Commonwealth Agricultural Bureaux, Kew, England.

Engelhard, A. W. 1993. Fungal leaf spots. Pages 228-229 in: Geraniums IV. J. W. White, ed. Ball Publishing, Geneva, IL.

Munnecke, D. E. 1956. Development and production of pathogen-free geranium propagative material. Pages 93-95 in: Plant Dis. Rep. Suppl. 238.

Alternaria Leaf Spot and Blight of Poinsettia

Alternaria leaf spot of the poinsettia, *Euphorbia pulcherrima* Willd. ex Klotzsch, has been reported from Hawaii and Florida in the United States and from Egypt.

Symptoms

The symptoms of Alternaria blight on poinsettia are similar to those of Phytophthora blight, bacterial canker, and scab. On bracts, purplish black spots, approximately 0.5 mm in diameter when first visible, develop into elliptical lesions (2–4 × 4–7 mm) and then into irregular, brown lesions as large as 8 cm in diameter with tan centers and purplish black borders. Angular or irregular lesions on inoculated leaves are dark brown with tan centers and range from 1 to 20 mm in diameter. Chlorotic halos often border lesions; severely infected leaves may become chlorotic and abscise (Plate 3). Narrow, dark lesions along leaf veins may also occur on partially expanded leaves, leading to leaf distortion. Stem lesions (3 × 8 mm) (Plate 4) are elongated and sunken. They are tan to brown with dark borders and may girdle the stem. Infection of the cyathia may also occur.

Causal Organism

Alternaria euphorbiicola E. Simmons & Engelhard has smooth, pale, tan conidia with often-lengthened apical cells (pseudorostra) that are not distinctly separable from the rest of the spore. Conidia are long-ovoid to long-ellipsoid and solitary or in chains of four to five or more. Each pseudorostrum functions as a conidiophore. Conidia have transverse, well-defined septa; the longitudinal septa are less distinct. The conidia of poinsettia isolates are 35–60 × 10–20 μm with a pseudorostrum of 8–12 μm and three to eight transverse septa and one to four longitudinal septa in the widest cells. Sporulation occurs on V8 juice, hay decoction, and potato-carrot agars at 20°C under a cycle of 8 hr of cool white fluorescent light and 16 hr of darkness.

Host Range and Epidemiology

Susceptibility varies among poinsettia cultivars. V-14 cultivars (Glory, White, and Jingle Bells II) and Eckespoint C-1 Red are highly susceptible; V-10 Amy has intermediate susceptibility; and Annette Hegg cultivars show resistance. Annette Hegg Dark Red, Top White, Brilliant Diamond, and Hot Pink develop only tiny leaf spots with tan centers. In Florida, the disease is most severe on outdoor poinsettia crops. The disease is not as much of a problem in greenhouses, where the foliage can be kept dry through careful environmental regulation.

Management

Cultural practices that minimize leaf wetness duration will help control this disease. Diseased plants and fallen leaf debris should be removed from the greenhouse or growing area. Poinsettia cultivars that are relatively resistant to *A. euphorbiicola* should be grown. Chlorothalonil and iprodione are registered in the United States for use against Alternaria blight on poinsettia. Combinations of fungicides are needed for integrated control of scab and Alternaria blight on susceptible poinsettia cultivars grown where both of these diseases occur.

Selected References

Engelhard, A. W. 1986. Poinsettia cultivar susceptibility to Alternaria blight. Biol. Cult. Tests Control Plant Dis. 1:69.

Engelhard, A. W., and Jones, J. B. 1985. A leaf, petiole and stem lesion disease of poinsettia incited by *Alternaria* sp. (Abstr.) Phytopathology 75:1306.

Simmons, E. G. 1986. *Alternaria* themes and variations (14–16). Mycotaxon 25:195-202.

Yoshimura, M. A., Uchida, J. Y., and Aragaki, M. 1986. Etiology and control of Alternaria blight of poinsettia. Plant Dis. 70:73-75.

Alternaria Leaf Spot of Cineraria

Alternaria leaf spot has been reported on cineraria, *Pericallis × hybrida* R. Nordenstam, grown in Great Britain, Europe, and the United States. This disease is not known to be important on cineraria crops in the United States, but its symptoms could easily be confused with leaf spotting produced by impatiens necrotic spot tospovirus and thus be misdiagnosed.

Symptoms

Leaf spots are initially dark, water soaked, and a few millimeters in diameter (Plate 5). Spots mature to a reddish or olivaceous brown color and may enlarge to 1 cm, sometimes coalescing to kill large portions of the leaves. The centers of the leaf spots may turn gray. Dark lesions may also develop on petioles.

Causal Organism

Alternaria cinerariae S. Hori & Enjoji (syn. *A. senecionis* Neergaard) has pale to midolivaceous brown, smooth conidiophores that measure up to 150 × 5–8 μm. The obpyriform to obclavate conidia (50–140 × 15–40 μm) are often in short chains and are rostrate, golden brown, and smooth. The conidia have three to 10 transverse and several longitudinal septa and are constricted at the septa. The beak is up to 80 μm long and 6–9 μm thick.

Host Range and Epidemiology

A. cinerariae is known to affect only cineraria. From studies of other Alternaria leaf spot diseases, it would be expected that an extended period of leaf wetness would be required for spore germination and infection and that the pathogen might be introduced into a greenhouse on seeds of the host.

Management

Infected leaves should be removed, and the length of time that the foliage remains wet should be minimized. Plants should be watered early in the day, and good air circulation should be provided. The development of condensation on leaf surfaces can be avoided by providing heat and ventilation at sunset. Iprodione is labeled for control of Alternaria leaf spot on cineraria in the United States.

Selected Reference

Cooper, A. J. 1956. The influence of cultural conditions on the development of Alternaria leaf spot of cinerarias. J. Hortic. Sci. 31:229-233.

Bipolaris Stem Rot and Leaf Spot of Holiday Cacti

Bipolaris stem rot and leaf spot of Easter cactus (*Hatiora gaertneri* (Regel) Barthlott) was first reported from Florida in 1980. Thanksgiving cactus (*Schlumbergera truncata* (Haw.) Moran) in the same nursery was not affected. In 1989, lesions and abscission of cladophylls of *S. truncata* in a California nursery were attributed to the same fungus.

Symptoms

Dark, water-soaked, mushy stem lesions appear both above and below the soil line on Easter cactus. Lesions develop a dark brown, felty appearance from dense, furry sporulation. Lesions as small as 2 mm result in cladophyll abscission (Plate 6). On *S. truncata*, dark, circular lesions accompanied by abscission develop on the cladophylls (Plate 7).

Causal Organism

Bipolaris cactivora (Petr.) Alcorn (syns. *Drechslera cactivora* (Petr.) M. B. Ellis and *Helminthosporium cactivorum* Petr.) is the fungus responsible for the stem rot and leaf shattering symptoms on holiday cactus (Plate 8). The fungus has straight or flexuous, caespitose conidiophores, which are sometimes swollen at the base and are usually swollen and irregularly lobed at the apex. The conidiophores are light to medium golden brown and up to 250 μm long and 4–6 μm wide, except at the swollen tips. Conidia are also light to medium golden brown and have two to four pseudosepta. They are 30–65 μm long, 9–12 μm at the widest point, and straight, fusiform, ellipsoidal, or obclavate.

Host Range and Epidemiology

B. cactivora has been reported from Florida, California, and Europe on cactus genera, including *Cereus*, *Echinopsis*, and *Mammillaria*. In inoculation trials, both wounded and unwounded *H. gaertneri* developed symptoms when sprayed with a spore suspension, although wounded plants developed symptoms more quickly (4–7 days vs. 14 days). Cladophyll abscission occurred as soon as 2 days after the appearance of lesions. Only in one test out of four did *B. cactivora* produce symptoms in *S. truncata*. Infection of *S. truncata* has not been observed under natural conditions in Florida but has been reported from California. Infection in *S. truncata* inoculation trials in California required wounding for disease development. The cultivars Rita and Majestic were the least susceptible, Annette was intermediate, and Maria was the most susceptible.

Management

Infected plants should be discarded, Easter cacti should be isolated from other cactus hosts, and media and pots should be steam pasteurized before reuse if there has been a disease outbreak. Chlorothalonil is registered in the United States for control of Bipolaris leaf spot on holiday cactus.

Selected References

Chase, A. R. 1982. Stem rot and shattering of Easter cactus caused by *Drechslera cactivora*. Plant Dis. 66:602-603.
Durbin, R. D., Davis, L. H., and Baker, K. F. 1955. A Helminthosporium stem rot of cacti. Phytopathology 45:509-512.
Raabe, R. D. 1989. Drechslera cladophyll blight of Christmas cactus. (Abstr.) Phytopathology 79.1218.

Botrytis Blight

Botrytis cinerea has a worldwide distribution and is ubiquitous in greenhouses. Botrytis blight is one of the most common diseases of greenhouse crops.

Symptoms

B. cinerea causes a range of symptoms including spots and blight on leaf or petal tissues (Plate 9), crown rot (Plates 10 and 11), stem cankers (Plates 12–15), cutting rot, and damping-off. Storage tissues such as roots, corms, or rhizomes are also susceptible. Wounded or senescent tissues are especially susceptible to invasion, but healthy tissues may also become colonized. Lesions caused by *B. cinerea* are often identified in the field by the characteristic gray, fuzzy sporulation (Fig. 1). However, spores develop only under humid conditions. Leaf lesions often develop a zonate pattern. Flower petals may have tiny flecks of discoloration or become completely blighted. Stems may die back from cutting wounds or develop tan to brown cankers that originate at the bases of petioles of blighted leaves.

Poinsettia (*Euphorbia pulcherrima* Willd. ex Klotzsch) is susceptible at all stages of production. During propagation, leaves may become blighted and cuttings may rot. On potted poinsettias, tan to brown lesions form on leaves, stems, and bracts. Extensive tan cankers can form on stems when *B. cinerea* enters via blighted petioles or shoots (Plate 16). Lesions begin at the margins of bracts, turning darker as they expand (Plates 17 and 18). Latex may exude from the undersides of lesions. Sporulation develops on the necrotic areas under humid conditions (Plate 19).

Cyclamen (*Cyclamen persicum* Mill.) petals, petioles, and leaves are susceptible to *B. cinerea*. On petals, the initial flecks enlarge to 1–4 mm, progressing from water-soaked lesions to tan, necrotic spots. Colored petals show rings of intensified color around the lesions. The petioles and flower stalks developing beneath the canopy may become infected and rot, causing portions of the plant to collapse. Subsequently, the infected petioles and pedicels typically become covered with grayish brown *B. cinerea* sporulation. Symptoms of Fusarium wilt or bacterial soft rot of cyclamen are quite similar; however, these diseases cause characteristic vascular discoloration and a soft, mushy rot of the corm, respectively.

Exacum (*Exacum affine* Balf. f.) is particularly susceptible to Botrytis blight. Young seedlings are susceptible to damping-off. A mature plant, which suddenly wilts, may have a reddish brown Botrytis canker at the base of the stem, originating at a node where leaves were buried at transplanting (Plate 20). This symptom may be confused with the basal canker produced by *Fusarium solani* (*Nectria haematococca*) and with the black stem canker caused by impatiens necrotic spot tospovirus (INSV). Initially, a stem canker on exacum caused by *B. cinerea* has a water-soaked appearance and may be bordered by a white halo. Both *B. cinerea* and INSV cause pale tan cankers on the secondary stems. Leaves infected with *B. cinerea* develop tan spots approximately 3 mm in diameter or bleaching or necrosis of the leaf margin. Lesions may coalesce to form large, tan, zonate, membranous lesions. Infected flowers exhibit water-soaked flecking on the petals or may become completely colonized and then collapse and turn brown. Flower blight frequently leads to colonization of the pedicel and stem.

On African daisy (*Gerbera jamesonii* H. Bolus ex Adlam), *B. cinerea* can cause damping-off, spotting and blighting of leaves and flowers, and crown rot. Leaves develop zonate lesions, and flower petals show tan spots and tip necrosis or are entirely blighted. *B. cinerea* may be seedborne in African daisy.

Geraniums (*Pelargonium* spp.) are very susceptible to *B. cinerea*. Annual losses for 1985 were estimated at $5.1–7.6 million worldwide. Losses are particularly high in vegetatively propagated *P. × hortorum* L. H. Bailey. Stock plants are grown close together and have multiple branches and a dense canopy that is created with growth regulators in order to maximize cutting productivity. Consequently, lower leaves of stock plants typically senesce and become colonized by *B. cinerea*, resulting in a massive buildup of inoculum. Stock plants are wounded when cuttings are harvested, often result-

ing in stem blight, which progresses down toward the main stem (Plate 21). *B. cinerea* infections may also develop at the cutting bases or at leaf scars during propagation under mist. Flower blight is common. *P. × hortorum* grown from seed is also susceptible to *B. cinerea*. Brown, zonate leaf lesions are often initiated after petals drop from the flowers or from a hanging basket crop above (Plate 22). Geraniums sometimes show ghost spots indicative of *Botrytis* lesions whose development has been arrested.

Causal Organism

Botrytis cinerea Pers.:Fr. is a hyphomycete with straight, brown, septate conidiophores, which are simple or, more commonly, alternately branched. The conidiophores bear botryose clusters of hyaline conidia, which appear grayish brown in mass (Plate 23). Conidia are budded off from a swollen, sporogenous cell at the tip of the conidiophore. The conidia (8–14 × 6–9 μm) are ellipsoid to ovoid. *Botrytis* spp. are pleomorphic: the conidial anamorph is in the form-genus *Botrytis* Pers.:Fr., the microconidial anamorph is *Myrioconium* H. Sydow, and the sclerotial anamorph is a *Sclerotium* Tode.

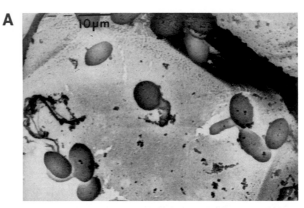

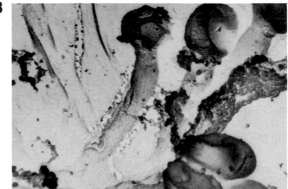

Fig. 2. Infection of a geranium cutting wound by *Botrytis cinerea*. **A,** Spore germination 3 hr after inoculation (1,000×); **B,** germ tube elongation 6 hr after inoculation (3,200×); and **C,** tissue invasion 12 hr after inoculation (100×). (Courtesy M. K. Hausbeck)

Fig. 1. *Botrytis cinerea* sporulation on cankered hydrangea stem tissue. (Courtesy J. A. Matteoni)

The teleomorph for *B. cinerea* is *Botryotinia fuckeliana* (de Bary) Whetzel, which is only rarely observed forming apothecia on sclerotia.

Colonies of *B. cinerea* on potato-dextrose agar are at first off-white and become gray to brown with the development of spores. The optimum temperature for growth is reported to be 24–28°C, but some growth occurs from 0 to 35°C. Sclerotia are black, hard, tightly appressed to the substrate, irregular in size and shape, and 1–15 mm long or confluent. Sclerotia have a dark, pseudoparenchymatous rind of nearly isometric cells approximately 5–10 μm in diameter enclosing a medulla of tightly knit, hyaline, thick-walled hyphae.

Host Range and Epidemiology

B. cinerea has a very broad host range, especially among dicots, and can cause "gray mold" disease within a wide range of temperatures. All of the potted plants covered in this compendium are susceptible to Botrytis blight.

Infection may occur directly or through natural openings or wounds (Fig. 2) by means of conidial germ tubes or by hyphal growth from colonized dead plant parts or organic debris that contacts healthy tissue. Infection by *B. cinerea* is stimulated by nutrient depletion of leaves and pollen deposition. Fruits and flowers of plants are generally more susceptible to infection than healthy, nonsenescent leaves. Bracts and flowers of poinsettias are particularly prone to infection by *B. cinerea* (Plate 18).

Conidia are released by a hygroscopic mechanism in association with a rapid change in relative humidity and require air currents or splashing water for dispersal within the greenhouse. In geranium stock-plant production areas, peak concentrations of *Botrytis* conidia have been associated with worker activity (Fig. 3). Harvesting cuttings, spraying pesticides, cleaning plants, and even drip-tube watering increases the number of *B. cinerea* conidia in the greenhouse atmosphere. Releases of 504–1,297 conidia per cubic meter per hour have been recorded during *P. × hortorum* cutting harvest, a time at which both cuttings and stock plants are freshly wounded.

Once dispersed, conidia can germinate in a water film that contains solutes; the more conidia in a water droplet, the greater the likelihood of an aggressive infection. African daisy flowers kept at 18–25°C and 100% relative humidity for only 5 hr develop small, necrotic lesions from single-conidium infections of the ray florets encompassing one to several epidermal cells. Exposure of open African daisy flowers to conidia and subsequent periods of high relative humidity

during shipment and storage will lead to the formation of small, necrotic lesions of the florets. Even after having been maintained under dry conditions for as long as 14 months, *B. cinerea* conidia retain their ability to infect African daisy florets once a film of moisture is available. Germination and lesion development on African daisy have been observed at 4–25°C, but not at 30°C. Poinsettias grown at constant 10°C show more Botrytis blight incidence than those grown at various 10°C night and 17°C day split-night temperature regimes or those grown at constant 17°C. Presumably, this is an indirect effect of higher relative humidity resulting from lower temperatures rather than a direct temperature effect.

Cultural practices often create opportunities for *B. cinerea* infections. For example, making cutting wounds on stock plants and stripping the lower leaves from cuttings facilitate infection. When crops of flowering plants are grown in hanging baskets over a crop susceptible to Botrytis blight, the fallen petals may serve as an energy source for the fungus and facilitate infection of adjacent healthy tissue.

Sclerotia formed in colonized plant tissues are important for long-term survival of *B. cinerea* in soil; under appropriate conditions, they will germinate and produce conidial (or, possibly, ascospore) inoculum.

Management

Control of *B. cinerea* is challenging because of its abilities to survive as a saprophyte, rapidly invade host tissues, and quickly produce abundant conidia that are easily distributed by air currents. Sanitation alone is not sufficient for minimizing Botrytis blight losses in the greenhouse. Prevention should be the main focus of a Botrytis blight management program. An integrated strategy combining environmental management, cultural practices, and fungicides will most effectively manage this omnipresent threat in the greenhouse.

Concerns about potential injury from fungicides, unsightly residues, limited registered fungicides, and the development of fungicide resistance make environmental management the obvious first line of defense against Botrytis blight. Effective control requires careful attention to managing leaf wetness duration and relative humidity. Providing ventilation and heat at sunset will help drive moisture-laden air out of the greenhouse, reducing the opportunities for infection during the night. Several cycles may be necessary. Adequate spacing between plants should be maintained, and open-mesh benching and horizontal airflow systems that improve air circulation in the greenhouse should be used.

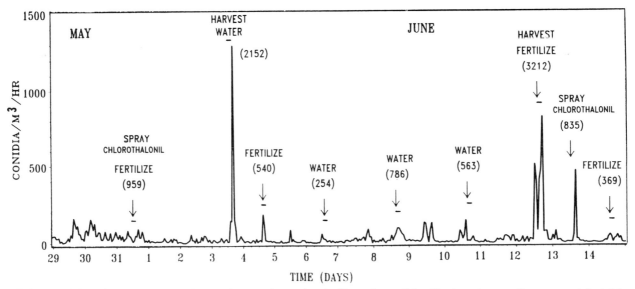

Fig. 3. Release of *Botrytis cinerea* spores in geranium stock correlated with worker activity. Numbers in parentheses are daily total spore counts. (Reprinted, by permission, from Hausbeck and Pennypacker, 1991)

For optimum and efficient control of Botrytis blight on geraniums, it has been recommended that growers regulate the environment and apply fungicide treatments at times when inoculum is high and wounds are available. Lowering relative humidity within a geranium canopy to less than 93% and controlling the occurrence and duration of free moisture on foliage are important environmental modifications.

Chemical control has been complicated by the appearance of widespread benzimidazole (including benomyl and thiophanate-methyl) resistance and a less pervasive dicarboximide (iprodione and vinclozolin) resistance in populations of *B. cinerea* in the greenhouse flower industry in Europe and North America. Insensitive strains are exchanged among greenhouses along with the exchange of plant material. After a greenhouse *B. cinerea* population acquires benzimidazole or dicarboximide resistance, it appears that the trait remains within the population after the fungicide is no longer being applied.

Benzimidazole and dicarboximide fungicides are useful only on nonresistant strains of *B. cinerea*. Other fungicides registered in the United States for control of Botrytis in greenhouses, including chlorothalonil, copper hydroxide, copper sulfate pentahydrate, and mancozeb, are generally more effective against *B. cinerea*. The use of tank mixtures that include a single-site and a multisite fungicide at reduced dosages has been proposed to avoid control failure caused by fungicide resistance. Product labels provide information about approved uses and tank mixtures on flowering potted plants.

Fungicide phytotoxicity may present a particular problem for growers attempting *Botrytis* control on crops during the finishing stages. Chlorothalonil smoke dust has been reported to injure poinsettia bracts and the flowers of several potted plants. Injury to poinsettias is often cultivar specific. High temperature and moisture on plant surfaces contribute to chlorothalonil smoke dust injury. Of several fungicides tested on exacum, only chlorothalonil gave effective control of *B. cinerea* with no phytotoxic effects.

Sprays of *Trichoderma harzianum* Rifai, a bioantagonistic fungus, have shown some efficacy against *Botrytis* on cyclamen, but the effect is inferior to control with iprodione. More widespread occurrence and recognition of greenhouse strains of *B. cinerea* resistant to some of the currently effective fungicides will encourage the development of biocontrol agents in the future.

Cultivars often vary remarkably in their susceptibility to *Botrytis*, but major gene resistance has not been identified for any plant species.

Selected References

Hausbeck, M. K., and Pennypacker, S. P. 1991. Influence of grower activity and disease incidence on concentrations of airborne conidia of *Botrytis cinerea* among geranium stock plants. Plant Dis. 75:798-803.

Jarvis, W. R. 1980. Epidemiology. Pages 219-250 in: The Biology of *Botrytis*. J. R. Coley-Smith, K. Verhoeff, and W. R. Jarvis, eds. Academic Press, London.

Jarvis, W. R. 1992. Managing Diseases in Greenhouse Crops. American Phytopathological Society, St. Paul, MN.

McCain, A. H. 1983. Gray mold control of cyclamen using *Trichoderma harzianum* and reduced rates of vinclozolin. Pages 1-2 in: Calif. Plant Pathol. 63.

Moorman, G. W., and Lease, R. J. 1992. Benzimidazole- and dicarboximide-resistant *Botrytis cinerea* from Pennsylvania greenhouses. Plant Dis. 76:477-480.

Moorman, G. W., and Lease, R. J. 1992. Residual efficacy of fungicides used in the management of *Botrytis cinerea* on greenhouse-grown geraniums. Plant Dis. 76:374-376.

Orlikowski, L., Hetman, J., and Tjia, B. 1974. Control of seed-borne *Botrytis cinerea* (Pers. ex Fr.) on *Gerbera jamesonii* Bolus.

HortScience 9:239-240.

Pappas, A. C. 1982. Inadequate control of grey mould on cyclamen by dicarboximide fungicides in Greece. Z. Pflanzenkrankh. Pflanzenschutz 89:52-58.

Tompkins, C. M., and Hansen, H. N. 1948. Cyclamen petal spot caused by *Botrytis cinerea*, and its control. Phytopathology 38:114-117.

Trolinger, J. C., and Strider, D. L. 1984. Botrytis blight of *Exacum affine* and its control. Phytopathology 74:1181-1188.

Trolinger, J., and Strider, D. L. 1985. Botrytis diseases. Pages 17-101 in: Diseases of Floral Crops. D. L. Strider, ed. Praeger, New York.

Vali, R. J., and Moorman, G. W. 1992. Influence of selected fungicide regimes on frequency of dicarboximide-resistant and dicarboximide-sensitive strains of *Botrytis cinerea*. Plant Dis. 76:919-924.

Cercospora Leaf Spot of Geranium

Cercospora leaf spot of *Pelargonium* spp. (geraniums) has been reported from the eastern United States, Canada, and the Philippines. It most commonly occurs in the southeastern United States but has not been significant in geranium production. Other Cercospora leaf spots (Table 4) are occasionally encountered on flowering potted plants, including *Hibiscus rosa-sinensis* L. and *Hydrangea macrophylla* (Thunb.) Ser. (Plates 24 and 25).

Symptoms

Leaf lesions on the florist's geranium, *Pelargonium* × *hortorum* L. H. Bailey, are at first pale green, sunken spots, 1–3

TABLE 4. *Cercospora* spp. that Cause Leaf Spots
on Flowering Potted Plants

Host	*Cercospora* sp.
Clerodendrum thomsoniae (bleeding heart vine)	*C. apii* f. sp. *clerodendri*
Dahlia hybrids	*Cercospora* sp.
Dicentra spectabilis (bleeding heart)	*Cercospora* sp.
Euphorbia pulcherrima (poinsettia)	*C. pulcherrimae*
Eustoma grandiflorum (lisianthus)	*C. eustomae*
Gardenia augusta	*Cercospora* sp.
Gerbera jamesonii (African daisy)	*C. gerberae*
Hibiscus rosa-sinensis	*C. malayensis*
Hydrangea macrophylla	*C. hydrangeae*
Kalanchoe blossfeldiana	*Cercospora* sp.
Pelargonium spp. (geranium)	*C. brunkii*
Schlumbergera truncata (Thanksgiving cactus)	*Cercospora* sp.
Verbena × *hybrida*	*Cercospora* sp.

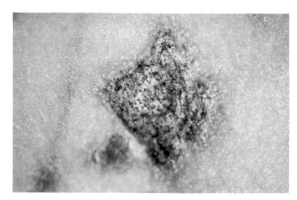

Fig. 4. Cercospora leaf spot on florist's geranium. (Courtesy M. Gleason)

mm in diameter. With time, the spots turn gray, and the accumulation of spores causes lesions to darken and appear to have raised centers (Fig. 4). When lesions coalesce, V-shaped necrotic areas develop, similar in appearance to those caused by bacterial blight. Chlorosis may develop in the vicinity of lesions. Heavily infected leaves abscise. Infection may be confused with *Pseudomonas cichorii* leaf spots, but those caused by *Cercospora brunkii* are usually smaller, have more chlorosis, and are tan to brown, while the bacterial leaf spots are brown to black and usually have water-soaked margins.

Causal Organism

Cercospora brunkii Ellis & B. T. Galloway (syn. *C. pelargonii*) is one of four *Cercospora* species described by Chupp on Geraniaceae and the only one of the four reported pathogenic to *Pelargonium*. The fungus produces hyaline, mostly acicular conidia (50–150 × 2.5–5 μm) with truncated bases. Conidiophores (50–200 × 4–5.5 μm) occur in fascicles of two to 13 stalks.

Host Range and Epidemiology

C. brunkii affects both *Pelargonium* and *Geranium*. Both *P. peltatum* (L.) L'Hér. (ivy geranium) and *P.* × *hortorum* are susceptible. Hybrid seed geraniums have not been screened for resistance, but the cultivar Showgirl has been reported to be somewhat resistant. Environmental conditions governing disease development have not been studied, but the pathogen is easily wind disseminated and requires a period of leaf wetness to infect.

Management

Prolonged periods of leaf wetness should be avoided. Chlorothalonil and thiophanate-methyl are labeled for control of Cercospora leaf spot on geranium in the United States.

Selected Reference

Chupp, C. 1953. A monograph of the genus *Cercospora*. Cornell University Press, Ithaca, NY.

Choanephora Wet Rot of Poinsettia

Choanephora cucurbitarum causes a soft, wet rot of poinsettias (*Euphorbia pulcherrima* Willd. ex Klotzsch). In Florida, the disease is most likely to occur during August, September, or October and may affect plants at all stages of production.

Symptoms

The symptoms of Choanephora wet rot resemble those caused by *Rhizopus stolonifer*. Leaves, petioles, and stems exhibit a soft, mushy decay. Green stems may droop before they collapse. Under humid conditions, young plants are entirely destroyed, while only a portion of an older plant may develop symptoms. *C. cucurbitarum* sporulates on diseased stems or other infected plant parts, producing characteristic whiskerlike sporangiophores (Plate 26).

Causal Organism

C. cucurbitarum (Berk. & Ravenel) Thaxt. (Choanephoraceae, Mucorales) produces long (up to 10 mm), aseptate sporangiophores, which are visible to the naked eye and create a whiskered look on infected plant parts. These sporangiophores bear compound heads of six or more clusters of indehiscent, unisporous sporangiola (Fig. 5). The pale brown,

Fig. 5. Sporangiophore of *Choanephora cucubitarum*. (Courtesy A. W. Engelhard)

reddish brown, or brown sporangiola are ellipsoidal and usually striate. They often have several hyaline appendages at either end as well as a small pedicel marking the end of attachment. *Rhizopus*, in contrast, has smooth, rounded spores, all of which are formed around a single, bulbous tip.

Host Range and Epidemiology

C. cucurbitarum also causes rot of certain fruits and vegetables and blighting of flowers of *Hibiscus rosa-sinensis* L. and *Petunia* × *hybrida* Vilm. *C. infundibulifera* (Curr.) Sacc. has been reported to cause a wet rot on young stems and flowers of hibiscus. The disease develops rapidly under warm, humid conditions. Observations of the fungus on *Capsicum annuum* L. (bell pepper) and *H. coccineus* (Medic.) Walt. (swamp hibiscus) indicate that sporulation ceases at temperatures below 14°C.

Management

Plant spacing to promote good air circulation, avoidance of prolonged leaf wetness, and removal of diseased plants have been suggested for management of this disease.

Selected References

Engelhard, A. W. 1987. Choanephora wet rot of poinsettia. Inst. Food Agric. Sci. Univ. Fla. GCREC Res. Rep. BRA 1987-23.
Kirk, P. M. 1984. A Monograph of the Choanephoraceae. Mycological Paper, no. 152. Commonwealth Mycological Institute, Kew, England.

Anemone Leaf Curl Caused by *Colletotrichum* sp.

Anemone leaf curl is an anthracnose disease that has caused major crop losses of anemones (*Anemone coronaria* L.) grown outdoors. Leaf curl was first reported from Australia in 1976, from Holland in 1977, and from Britain, France, and Italy in 1978. The disease has not been formally reported in the United States, but it has been observed in New York. The causal fungus is easily transported over long distances via seed or rhizomes.

Symptoms

The leaf margins curl down, and the petioles and pedicels become extraordinarily twisted so that the plant appears to be injured by a growth-regulator herbicide. Curling may also result from downy mildew or rust, but the leaf curl disease is unique in the severity of the growth distortion that it causes. Dieback of new shoots and stem splitting may occur, and young leaves do not expand fully. Small, brown lesions that exhibit wet, orange spore masses under humid conditions may develop on petioles and pedicels (Plate 27).

Causal Organism

The causal fungus of anemone leaf curl is *Colletotrichum* sp. The pathogen was originally identified as *C. acutatum* J. H. Simmonds, but there is some question as to whether the fungus is actually *C. gloeosporioides* (Penz.) Penz. & Sacc. in Penz. Sutton describes *C. acutatum* as having straight, fusiform conidia (8.5–16.5 × 2.5–4 μm); not forming sclerotia in culture; having sparse, clavate to irregular mycelial appressoria (8.5–10 × 4.5–6 μm); causing pink pigmentation on potato-dextrose agar; and having a wide host range. The *C. gloeosporioides* group also has a wide host range. Colonies of *C. gloeosporioides* are described as variable, and sclerotia are occasionally present. *C. gloeosporioides* conidia (9–24 × 3–4.5 μm) are straight and obtuse at the apical end, and the mycelial appressoria (6–20 × 4–12 μm) are clavate to irregular.

Host Range and Epidemiology

The extent of the host range of the pathogen that occurs on *A. coronaria* is as yet unknown. In Australia, *C. acutatum* is known to be a pathogen of *Ranunculus asiaticus* L. and some fruit and vegetable crops. The anemone leaf curl pathogen may be seedborne. It is most commonly distributed on symptomless rhizomes, where it is associated with the debris from the previous season's stalks. The fungus was shown to survive in soil for at least 18 months, but there was no disease recurrence when anemones were replanted into fields 2 or 3 years after a severely infected crop had been incorporated into the soil. Warm, moist conditions favor sporulation and disease spread; the optimum temperature for the fungus in vitro is 25°C. An improved medium for isolating the pathogen from soil with a cornmeal agar base and low pH to reduce contaminants has been developed.

Management

Fungicide sprays are reasonably effective at controlling anemone leaf curl when environmental conditions are not optimal for disease development. Thiophanate-methyl has been used for the control of this disease in Europe. All precautions should be taken to avoid introducing *Colletotrichum* sp. into the greenhouse on anemone seed or corms. For field-grown stock, a minimum 2-year rotation with a nonsusceptible host should be followed.

Selected References

Barker, I., and Pitt, D. 1987. Selective medium for the isolation from soil of the leaf curl pathogen of anemones. Trans. Br. Mycol. Soc. 88:553-555.

Barker, I., and Pitt, D. 1988. Detection of the leaf curl pathogen of anemones in corms by enzyme-linked immunosorbent assay (ELISA). Plant Pathol. 37:417-422.

Gullino, M. L., and Garibaldi, A. 1981. Results of experimental trials for controlling leaf curling of anemone caused by *Colletotrichum gloeosporioides*. Med. Fac. Landbouwwet. Rijksuniv. Gent 46:873-879.

O'Neill, T. O. 1983. Curl up and die. GC & HTJ 194:24-25.

Sutton, B. C. 1980. The Coelomycetes. Commonwealth Mycological Institute and Commonwealth Agricultural Bureaux, Kew, England.

Woodcock, T., and Johnston, L. 1980. Evaluation of fungicides as foliar treatments for the control of leaf and stem curling in anemones, 1979. Fungic. Nematicide Tests 38:169.

Cyclamen Anthracnose Caused by *Glomerella cingulata*

Cyclamen anthracnose has been widely observed in the United States and reported on *Cyclamen persicum* Mill. from Florida, Indiana, Massachusetts, Missouri, North Carolina, New Jersey, Ohio, Pennsylvania, Texas, and Virginia. It has also been observed in New York. The disease can severely disfigure both leaves and petals of cyclamen.

Symptoms

Cyclamen anthracnose causes small, round, brown leaf spots (Plate 28) that resemble the leaf spot symptoms caused by impatiens necrotic spot tospovirus (INSV). Leaf symptoms are particularly distinct when viewed from the underside of the leaf, where the spots are round, brown, and sunken (Fig. 6). Spots may be numerous and may coalesce to form large patches of necrosis. Leaves may be entirely blighted, and the young pedicels and unexpanded leaves beneath the canopy may also be killed (Plate 29). To distinguish these fungal leaf spots from those caused by INSV, the upper surfaces of the lesions can be examined closely for orange masses of sporulation oozing from pimplelike acervuli. Flowers will also develop brown lesions (Plate 30), which might be confused with Botrytis infection. However, acervuli will also develop on petal tissue, aiding in identification.

Causal Organism

Glomerella cingulata (Stoneman) Spauld. & H. Schrenk has black, ostiolate, beaked, obpyriform to subglobose perithecia, which are at least partially immersed in the plant tissue. There are hairs around the ostiole. The fungus has hyaline, unicellular, ellipsoid to fusiform ascospores less than 20 μm long, which are biseriate within the unitunicate, elongate to cylindrical asci. Periphyses and paraphyses are present within the perithecium. The smaller size of the perithecia and ascospores in *Glomerella* help to distinguish it from *Physalospora*.

The anamorph of *G. cingulata* is *Colletotrichum gloeosporioides* (Penz.) Penz. & Sacc. in Penz., which produces spores in acervular conidiomata that may contain setae (see Anemone Leaf Curl Caused by *Colletotrichum* sp.). *C. gloeosporioides* is very similar to *Cryptocline cyclaminis*, which causes another anthracnose disease on cyclamen (see Cyclamen Anthracnose Caused by *Cryptocline cyclaminis*).

Host Range and Epidemiology

G. cingulata causes anthracnose symptoms on a wide range of woody and herbaceous ornamentals (Table 5) and may also be a saprophyte on injured tissue. *G. cingulata* is a cosmopolitan species that can be introduced into the greenhouse on numerous flower or foliage crops. Spore inoculum could

Fig. 6. Leaf spots caused by *Colletotrichum gloeosporioides* on the underside of a cyclamen leaf.

TABLE 5. Anthracnose Diseases of Flowering Potted Plants

Host Plant	Pathogen	Disease
Begonia spp.	*Glomerella cingulata*[a]	Anthracnose
Bougainvillea spectabilis	*Colletotrichum* sp.	Anthracnose
Capsicum annuum	*G. cingulata*	Anthracnose
Catharanthus roseus (vinca)	*Colletotrichum dematium*	Twig blight
	Colletotrichum sp.	Anthracnose
Cyclamen persicum	*Cryptocline cyclaminis*	Anthracnose
	G. cingulata	Bud and leaf blight, leaf spot
Dahlia hybrids	*Colletotrichum coccodes*	Root and stem rot
	Diplodina dahliae	Anthracnose
Dicentra spectabilis (bleeding heart)	*Colletotrichum* sp.	Leaf spot
Euphorbia pulcherrima (poinsettia)	*G. cingulata*	Anthracnose
Gardenia augusta	*G. cingulata*	Anthracnose
Hibiscus rosa-sinensis	*G. cingulata*	Anthracnose, dieback
Hydrangea macrophylla	*Colletotrichum* sp.	Leaf spot
Pelargonium × *hortorum*	*Glomerella* sp.	Leaf spot
Primula Pruhonicensis hybrids (polyanthus)	*G. cingulata*	Anthracnose, leaf spot
Ranunculus asiaticus	*Colletotrichum acutatum*	Anthracnose
Schizanthus sp. (butterfly flower)	*Colletotrichum schizanthi*	Anthracnose
	Glomerella sp.	Leaf spot
Schlumbergera truncata (Thanksgiving cactus)	*G. cingulata*	Zonate leaf spot
Sinningia speciosa (gloxinia)	*Colletotrichum* sp.	Anthracnose
Solanum pseudocapsicum (Jerusalem cherry)	*Glomerella* sp.	Anthracnose
Verbena × *hybrida*	*Glomerella* sp.	Leaf spot

[a] The anamorph of *G. cingulata* is *Colletotrichum gloeosporioides*, which is frequently found on diseased tissue.

also conceivably originate from infected native or landscape plants and be introduced through open vents by wind-driven rain or insects. The fungus is spread by splashing water, and wet leaf surfaces allow infection. Symptoms ordinarily become apparent during the spring or summer when greenhouse conditions are warm and plants are frequently irrigated.

Management

Overhead irrigation should be used as infrequently as possible during the production cycle, and irrigating should be done early in the day to allow foliage to dry before nightfall. Subirrigation will minimize spread. Good air circulation should be maintained. Plants should be scouted for the appearance of leaf spot symptoms, particularly while plants are small and closely spaced. Fungicides used for control of cyclamen anthracnose in Europe include chlorothalonil and thiophanate-methyl.

Cyclamen Anthracnose
Caused by *Cryptocline cyclaminis*

Anthracnose caused by *Cryptocline cyclaminis* is considered to be the most serious disease of cyclamen (*Cyclamen persicum* Mill.) in Europe. It occurs in the United States as well but has been reported less frequently. This disease is distinct from the anthracnose on cyclamen caused by *Glomerella cingulata*, although the two diseases and the two pathogens can easily be confused.

Symptoms

The symptoms of this anthracnose include round, brown spots on leaves and stunting, deformity, and necrosis of the immature petioles and pedicels at the center of the plant (Plate 31). There is also a discoloration of the vascular system of infected petioles and a reddish discoloration inside the corm. The symptoms of *C. cyclaminis* on immature leaves are easily confused with those of plants injured by feeding of a tarsonemid mite (cyclamen mite). The presence of waxy, pale orange pink masses of spores on necrotic stems or leaves allows this disease to be distinguished from mite injury. Symptoms and the appearance of sporulation at low magni-

fication are similar for this anthracnose and that caused by *G. cingulata*.

Causal Organism

C. cyclaminis (Sibilia) von Arx is the revised name for the fungus previously known as *Gloeosporium cyclaminis*. Its hyaline to pale brown hyphae are 3–4 μm wide. The acervuli (100–150 μm in diameter) are at first covered by the cuticle and then erupt as small, white to pale orange pustules (Plate 32). The acervuli are often clustered together and are sometimes confluent, the pseudoparenchymatous stroma, 35–60 μm thick and partially immersed in the host tissue, is made up of isodiametric or slightly elongate, hyaline to pale brown cells. The conidia are formed successively from phialides on straight to flexuous, smooth-walled conidiophores (17–28 × 3–5 μm), each of which has a small, flared collarette at the apex. The unicellular conidia (12–16 × 4–6 μm) are hyaline, oblong to ellipsoidal, guttulate, and truncated at the base. The teleomorph for this genus is *Trochila*.

Host Range and Epidemiology

C. persicum is the only reported host of this pathogen. The members of the genus *Cryptocline* are facultative parasites and only occasionally cause leaf spots or stem lesions. Disease development is favored by high humidity and leaf wetness. The severity of infection increases with increasing temperatures of 14–26°C. The spores are produced in wet masses and disseminated by water splash and possibly by insects or workers' hands. Seed transmission is also likely.

Management

Since normal spring greenhouse conditions strongly favor this disease, careful attention to minimizing leaf wetness duration and high temperature, particularly during the initial stages of crop production when plants are closely spaced, is important for control. Removal of diseased plant debris is also important. Plants should be spaced to reduce splash dispersal. In Europe, *C. cyclaminis* has shown some resistance to benzimidazole fungicides.

Selected References

Krebs, E.-K. 1986. Brennfleckenkrankheit ist *Cryptocline cyclaminis*. Gb + Gw 86:1868-1869.

Krebs, E.-K. 1987. Brennflecken-Krankheit ist sicher bekämpfbar. (Anthracnose is definitely controllable.) Gb + Gw 87:1731-1733.

Rampanini, G. 1991. La coltivazione del ciclamino: Non tutto, ma quasi. Clamer Informa. Suppl. 9. Pentagono Editrice, Milan.

Corynespora Leaf Spot of *Catharanthus*, *Hydrangea*, Poinsettia, and *Saintpaulia*

Corynespora leaf and bract spot on *Euphorbia pulcherrima* Willd. ex Klotzsch (poinsettia) was first reported from Florida in 1984. It was reported on *Hydrangea macrophylla* (Thunb.) Ser. in Florida in 1960 and in 1966 was described as the most serious leaf spot disease on that crop. *Catharanthus roseus* (L.) G. Don (vinca) and *Saintpaulia ionantha* Wendl. (African violet) are also susceptible to leaf infection.

Symptoms

On poinsettia, large, irregularly shaped, brown lesions form on bracts and leaves (Plate 33 and Fig. 7) and strongly resemble those caused by *Botrytis cinerea* infections. They occur primarily at the tips and margins of leaves and may be as large as 3 cm in diameter. On hydrangea, symptoms range from tiny, reddish purple spots to larger (3–5 mm in diameter), circular lesions that have tan centers and reddish purple margins. Although Corynespora leaf spot of hydrangea might be confused with Cercospora leaf spot, Corynespora leaf spots are generally smaller and the spores of these two pathogens are easily distinguished by microscopic examination. *Corynespora cassiicola* also causes irregular, brown lesions on African violets (Plate 34). Leaf spots on vinca start as 1- to 2-mm necrotic lesions, often at the leaf margins, and enlarge to

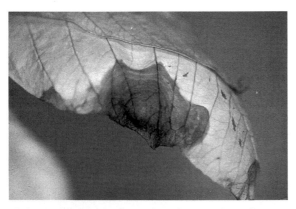

Fig. 7. Large bract lesion on poinsettia caused by *Corynespora cassiicola*. (Courtesy G. W. Simone)

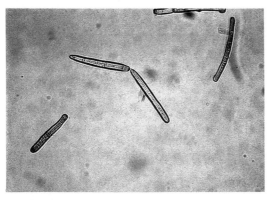

Fig. 8. Large, multiseptate spores of *Corynespora cassiicola*. (Courtesy G. W. Simone)

round to oblong, targetlike lesions (5.2–6.4 mm). On both vinca and azalea, infection causes premature leaf abscission.

Causal Organism

C. cassiicola (Berk. & M. A. Curtis) C. T. Wei (syn. *Helminthosporium cassiicola* Berk. & M. A. Curtis) is a hyphomycete with brown conidiophores (110–850 × 4–11 μm) that are usually unbranched and have up to nine cylindrical proliferations. The olivaceous brown conidia (40–220 × 9–22 μm) vary in shape. They are straight or curved, have four to 20 pseudosepta, and often exhibit a dark hilum at one end (Fig. 8). They are solitary or catenate on the erect conidiophores. On potato-dextrose agar, optimum growth occurs at 22–26°C. The aerial mycelium is medium gray to dark greenish gray, while the submerged mycelium is dark gray to black. The fungus sporulates moderately well in culture.

Host Range and Epidemiology

A variety of plants are hosts for *C. cassiicola*, including *Aeschynanthus pulcher* (Blume) G. Don (lipstick vine), *Aphelandra squarrosa* Nees (zebra plant), *C. roseus*, *Rhododendron obtusum* (Lindl.) Planch. (azalea), *Begonia*, *H. macrophylla*, *E. pulcherrima*, *S. ionantha*, and *Salvia splendens* Sellow ex Roem. & Schult. (scarlet sage), as well as vegetable and field crops. Most isolates show no host specialization. High moisture and humidity are conducive to infection. Wounding is necessary for infection on *Aphelandra* but is not usually required on other hosts. *C. cassiicola* is spread by airborne spores and is also seedborne. It has been observed to survive on plant debris for 2 years.

Management

Reduction of humidity and leaf wetness duration are important for preventing infection. Chlorothalonil sprays are registered in the United States for the control of Corynespora leaf spot on begonia, hydrangea, and poinsettia.

Selected References

Chase, A. R., and Simone, G. W. 1986. Corynespora bract spot of *Euphorbia pulcherrima* in Florida. Plant Dis. 70:1074.

Ellis, M. B., and Holliday, P. 1971. *Corynespora cassiicola*. Descriptions of Pathogenic Fungi and Bacteria, no. 303. Commonwealth Mycological Institute, Kew, England.

Holcomb, G. E., and Fuller, D. L. 1993. Corynespora leaf spot of poinsettia in Louisiana. Plant Dis. 77:537.

Sobers, E. K. 1966. A leaf spot disease of azalea and hydrangea caused by *Corynespora cassiicola*. Phytopathology 56:455-457.

White Smut of *Dahlia*

The white smut pathogen *Entyloma calendulae* f. *dahliae* occasionally causes leaf spots on *Dahlia* hybrids during greenhouse production. Growers may be slow to recognize white smut as a fungus disease, since microscopic examination is necessary to observe the distinctive ustilospores within the leaf tissue.

Symptoms

Dahlias with white smut disease show large (up to 1 cm in diameter), round, elliptical, or angular leaf spots (Plate 35 and Fig. 9). The spots change in color from pale to brown as they mature.

Causal Organism

E. calendulae (Oudem.) de Bary f. *dahliae* (Sydow) Viégas is commonly referred to as a white smut. The sori are embedded in the leaves (Fig. 10). The ustilospores (8–14 µm in diameter) are hyaline to pale yellow and globose to polygonal. The spores have a double-layered, smooth wall 1–3 µm wide. On the surfaces of the leaf spots, variably shaped, hyaline conidia (10–25 × 2–3.3 µm) are produced. *E. calendulae* f. *dahliae* can be cultured. The host range of this pathogen differs from that of *E. calendulae* f. *calendulae*, which is morphologically indistinguishable.

Host Range and Epidemiology

White smut is known to attack *D. coccinea* Cav., *D. pinnata* Cav., and cultivated hybrids and is distributed worldwide. Not all *Dahlia* spp. are hosts. The fungus can be soilborne or overwinter as ustilospores in plant debris. Although seed transmission has not been reported, the appearance of white smut in seed-propagated greenhouse crops suggests that the pathogen may be carried with seed. Within the greenhouse, spread can be accomplished by splashing conidia from plant to plant during overhead irrigation. Disease development is favored by high humidity. The mature ustilospores are viable for a long time under dry conditions.

Management

Removal of diseased plants from the growing area and wide pot spacing are useful for reducing disease impact on a dahlia crop. Lime and potassium applications have been helpful in some instances.

Selected Reference

Morduc, J. E. M. 1984. *Entyloma calendulae* f. *dahliae*. Descriptions of Pathogenic Fungi and Bacteria, no. 802. Commonwealth Mycological Institute, Kew, England.

Fig. 9. Haloed leaf spots on dahlia caused by a white smut, *Entyloma calendulae* f. *dahliae*.

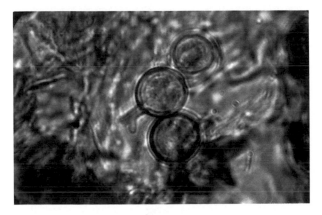

Fig. 10. Round ustilospores of a white smut inside leaf tissue.

Myrothecium Leaf Spot of *Begonia, Gardenia,* Gloxinia, and New Guinea Impatiens

Myrothecium leaf spot and canker were reported in the United States on *Gardenia augusta* (L.) Merrill in 1941. *Myrothecium roridum* is widely distributed and affects a wide range of hosts, usually causing minor leaf spots and sometimes shot holes on leaves. It may also occur on stem tissues, particularly at the soil line. During propagation, this disease may cause defoliation or extensive loss of cuttings.

Symptoms

On gardenia, circular and then irregular leaf spots develop to about 2.5 cm in diameter (Plate 36). The lesions are often zonate and extend across the leaf veins. The center of the spot may be a lighter shade of brown than the advancing margin, and the lesion may be darker in color on the upper side of the leaf than it is on the undersurface. Sporulation is more abundant on the lower leaf surface than on the upper surface. Distinctive dark green to black sporodochia (1–2 mm in diameter) rimmed with white hyphal tufts appear scattered across the lesion (Plate 37). Occasionally, the fruiting bodies develop in concentric rings. The fungal sporulation on the lesions clearly distinguishes Myrothecium leaf spots from those caused by *Mycosphaerella gardeniae* (Plate 38).

On other hosts, such as *Begonia* and New Guinea impatiens, leaf lesions are similar to those described for gardenia (Plate 39). In general, *M. roridum* infections occur at wounds such as those at the bases of cuttings or where leaves have been injured. On highly susceptible hosts, such as *Sinningia speciosa* (Lodd.) Hiern (gloxinia), infection occurs without apparent wounding (Plate 40). On foliage plants, leaf edges and tips are likely infection sites. Young, tissue-cultured plantlets are especially susceptible. Leaf lesions on New Guinea impatiens (Plate 39) are similar to those caused by impatiens necrotic spot tospovirus. When zonate leaf lesions on New Guinea impatiens occur, the undersides should be checked for *M. roridum* sporulation.

Causal Organism

M. roridum Tode:Fr. is a common, soil-dwelling saprophyte that is pathogenic to susceptible hosts under favorable environmental conditions (facultative parasite). It produces conidia in distinctive sporodochia, which are approximately 1–2 mm in diameter (Plate 37). The young spore mass is wet and greenish at first and then dries to a hard, black mass surrounded by a fringe of white hyphae. Conidia are produced on phialides arranged in whorls of two to five at the apex of conidiophores that are closely compacted into parallel rows. Conidia (4.5–10.8 × 1.3–2.7 µm, average 7.2 × 1.8 µm) are rod shaped with rounded ends, hyaline to pale olivaceous, and guttulate. The saprophyte *M. verrucaria* may be distinguished by its oval spores.

On potato-dextrose agar, *M. roridum* produces floccose, white colonies that are wrinkled, sometimes raised in the center, and pinkish on reverse. In culture, black sporodochia generally form within 10 days, often in concentric rings (Plate 41).

Host Range and Epidemiology

M. roridum has a wide host range, including *Begonia*, *G. augusta*, *S. speciosa*, New Guinea impatiens, and numerous foliage plants. The fungus grows over a wide range of pH values from 4.6 to 10.6. At pH 3.6, spore germination is greatly reduced. Optimum temperature for growth is 25°C, although growth occurs at 6–35°C. Optimum temperature for spore germination is 28°C.

Studies on Myrothecium leaf spot of *Dieffenbachia maculata* (Lodd.) G. Don have indicated that the optimum temperature for disease development is 21–27°C. Fertilizer rates above optimum for dieffenbachia quality increase disease incidence. Insecticidal soap at labeled rates has been observed to increase the severity of infection on the cultivar Perfection. The disease occurs under greenhouse conditions when relative humidity is high and moisture is most abundant, such as during propagation under mist.

High temperature (32°C) for as little as 4 hr per day completely suppressed Myrothecium symptom development on *D. maculata* 'Perfection' when plants were grown at 24°C for the rest of the day. Since constant 32°C allowed fungus growth in in vitro studies, it was postulated that the high temperatures were optimal for wound periderm formation and thus may have brought about control by altering potential infection sites for the fungus.

Management

Practicing careful sanitation, minimizing leaf wetness duration, avoiding unnecessary wounds, and promoting rapid wound periderm formation after cuttings have been taken help to manage Myrothecium leaf spot. If temperatures are anticipated to be between 21 and 27°C at the time of cutting harvest, fungicide should be applied to stock plants and cuttings of highly susceptible species. Chlorothalonil is registered in the United States for use on begonias and impatiens for the control of *M. roridum*.

Selected References

Fergus, C. L. 1957. *Myrothecium roridum* on gardenia. Mycologia 49:124-127.

Fitton, M., and Holliday, P. 1970. *Myrothecium roridum*. Descriptions of Pathogenic Fungi and Bacteria, no. 253. Commonwealth Mycological Institute, Kew, England.

Cyclamen Stunt
and Ramularia Leaf Spot
of *Cyclamen* and *Primula*

Ramularia cyclaminicola causes a variety of symptoms on cyclamen (*Cyclamen persicum* Mill.) collectively referred to as cyclamen stunt. Stunting, wilting, and leaf symptoms are all manifestations of infection by this fungal pathogen. The disease was for some time reported only from North America. It was much more common in the first half of the 20th century than it is today. A Ramularia leaf spot caused by a different species occasionally affects *Primula* spp. (primroses).

Symptoms

The older leaves of mature cyclamen may develop diffuse yellow areas and wilt when infected by *R. cyclaminicola*. A frosty growth of sporulation appears on the undersides of yellowed leaves nearest the soil surface. Brown spots with indefinite margins (0.5–5.0 mm in diameter) sometimes occur on the lower surfaces of leaves, and these spots may coalesce. Gray patches develop on the upper leaf surface opposite these spots, and then leaves dry and collapse. The entire plant is stunted, and the pedicels are shortened so that the flowers bloom beneath the leaf canopy. There is an associated reddish brown discoloration in the upper portion of the corm, sometimes extending into petioles, peduncles, and roots (Plate 42). In some cases, wilting and yellowing of leaves and corm discoloration have occurred without stunting. The disease may be confused with a physiological problem or with Fusarium or Phialophora wilts. The characteristic *Ramularia* sporulation on chlorotic leaves helps to distinguish this disease from others that cause similar symptoms.

Ramularia leaf spot on primulas also occurs on the older leaves. The spots are tan or brown and angular and may be accompanied by chlorosis (Plate 43). The white spore masses that develop on the undersides of lesions under humid conditions distinguish the disease from leaf spot caused by *Septoria primulae*.

Causal Organisms

R. cyclaminicola Trel. was mistakenly identified as a *Cladosporium* sp. when it was studied in culture rather than on the leaves of its host. *Ramularia* grows more slowly in culture than *Cladosporium*, although the dark green colony coloration and spore morphology of the two genera are similar. On host tissue, clusters of hyaline conidiophores project from stomata. Conidia are hyaline, cylindrical, and one to two celled; they usually occur in chains. On potato-dextrose agar, *R. cyclaminicola* colonies change from white to pale gray to gray green with whitish margins, reaching 7 mm in diameter over a 10-day period. After prolonged storage on agar media, cultures acquire a cottony, white mycelial growth, and pathogenicity is lost. *R. primulae* Thuem. is the cause of leaf spot on primulas.

Host Range and Epidemiology

R. cyclaminicola is known only on *C. persicum*. *R. primulae* is known to affect both *Primula* Pruhonicensis hybrids (polyanthus) and *P. malacoides* Franch. (fairy primrose). Since the Ramularia leaf spot of cyclamen was for many years reported only from North America and not from Germany or Holland where corms and seeds for production originated, it has been speculated that the pathogen originated on some native North American plant belonging to the Primulaceae. *R. cyclaminicola* may be transmitted as mycelium within the integument of seed and by movement of infested soil and airborne conidia. Conidia germinate in 15–20 hr on water agar. Disease development on cyclamen is slow, requiring 4 months or longer.

Management

Symptomatic plants should be removed and discarded. Seedlings of cyclamen and primrose should be grown away from older plants to prevent transfer of airborne inoculum. Soil used as a mix component should be steam pasteurized. Thiophanate-methyl is registered in the United States for control of diseases caused by *Ramularia* spp. on ornamentals.

Selected References

Baker, K. F., Dimock, A. W., and Davis, L. H. 1950. *Ramularia cyclaminicola* Trel., the cause of cyclamen stunt disease. Phytopathology 40:1027-1034.

Jacques, J. E. 1942. La dissemination d'un *Cladosporium*, cause du nanisme des cyclamens. Compt. Rend. Soc. Biol. Montreal Rev. Can. Biol. 1:685-686.

Massey, L. M., and Tilford, P. E. 1932. Cyclamen stunt. Phytopathology 22:19.

Rampanini, G. 1991. La coltivazione del ciclamino: Non tutto, ma quasi. Clamer Informa. Suppl. 9. Pentagono Editrice, Milan.

Smith, R. E. 1940. Diseases of flowers and other ornamentals. Pages 19-20 in: Calif. Agric. Ext. Circ. 118.

Tilford, P. E. 1932. Diseases of ornamental plants. Pages 31-32 in: Ohio Agric. Exp. Stn. Bull. 511.

Trelease, W. 1916. Two leaf-fungi of cyclamen. Trans. Ill. State Acad. Sci. 9:141-146.

Wagner, V. R. 1987. Ramularia leaf disease of cyclamen. Pages 3-4 in: Regul. Hortic. Penn. Dep. Agric. Bur. Plant Ind. Plant Pathol. Circ. 52.

Rhizopus Blight of Poinsettia, *Gerbera*, *Crossandra*, and *Catharanthus*

Rhizopus blight of *Gerbera jamesonii* H. Bolus ex Adlam (African daisy), *Crossandra infundibuliformis* (L.) Nees (crossandra), and *Euphorbia* spp. (including poinsettia) was reported during the 1980s in Florida. The pathogen is more commonly known as a bread mold or storage mold on ornamentals, soft fruits, and vegetables. Rhizopus blight has also been reported on *Catharanthus roseus* (L.) G. Don (vinca) and *Sinningia speciosa* (Lodd.) Hiern (gloxinia).

Symptoms

Under Florida growing conditions, flowers, leaves, and stems of poinsettia, crossandra, and African daisy are infected. The tissue is blighted, and webs of mycelium may develop over the dead tissue under humid conditions (Plates 44 and 45). The sporangia of the fungus appear to the eye as black specks on the white mycelial webs. Poinsettias may show symptoms during propagation or soon after potting. Rhizopus blight of gloxinia flowers has also been observed (Plate 46). In the northern United States as well as in Florida, *Rhizopus stolonifer* causes a stem rot of poinsettia. A dark, greasy-looking discoloration extends 4 inches or more up from the base of the stem. The hyphal growth is generally visible only inside the stem by microscopic examination. Potted poinsettias may suddenly wilt and collapse after infection (Plate 47) and exhibit symptoms that may be mistaken for bacterial soft rot. On vinca, *R. stolonifer*-infected tissues are water soaked and exhibit a soft rot. The infection leads to complete plant collapse (Plate 48), and sporulation may occur on the stem (Fig. 11).

Causal Organism

R. stolonifer (Ehrenb.:Fr.) Vuill. (syn. *Rhizopus nigricans* Ehrenb.) (Mucorales) has a worldwide distribution. It produces stolons that grow across the substrate and are anchored by rhizoids at each node (internodes may be as long as 1–3 cm). The hyphae are somewhat branched. The hemispheric sporangia (85–200 µm in diameter) (Fig. 12) are produced on long (1,000–2,000 µm) sporangiophores, which are usually clumped into groups of three to five or more. The brownish black sporangiospores (averaging 18 × 7.8 µm) produced within the sporangia are irregular, round or oval, and angular and show longitudinal striations. The extremely broad, coenocytic hyphae may be seen clearly in soft-rotted tissue stained with cotton blue in lactophenol. Colonies on potato-dextrose agar are cottony white initially and turn brownish black with age. The fungus sporulates abundantly in culture and on colonized plant parts under humid conditions.

Host Range and Epidemiology

R. stolonifer was reported as a pathogen of *Euphorbia* spp. previous to its documentation as a pathogen of *C. infundibuliformis*, *G. jamesonii*, and *C. roseus*. It is particularly aggressive on *E. milii* Des Moul. (crown-of-thorns). The fungus affects a wide range of nonornamental hosts as well, most commonly causing a soft rot of fruits. Annette Hegg poinsettia (*E. pulcherrima* Willd. ex Klotzsch) cultivars are reported to be less susceptible than V-14 Glory and V-14 White.

The spores of *R. stolonifer* are ubiquitous and can survive periods of desiccation. The pathogen grows rapidly at 21–32°C, and disease develops only at relative humidities greater than 75%. *R. stolonifer* may either colonize wounds or act as an opportunistic pathogen, affecting crops growing under stressful conditions. In some plants, infection begins in the flowers and extends down the stem to kill the plant. In Florida, disease outbreaks on flowering plants are most likely to occur during the fall, and wounding is not a prerequisite to infection. Vinca infection in Great Britain has been observed only on wounded plants during periods of high temperature and high humidity.

Management

The crop should be maintained under optimal cultural conditions, because outbreaks of this disease have generally followed periods of stress. It is particularly important to maintain ideal conditions for prompt callusing and rooting of cuttings, since the cutting wound is a major point of entry for the fungus. Infected plant debris should be removed.

Selected References

Engelhard, A. W. 1982. Poinsettia diseases and their control. Inst. Food Agric. Sci. Univ. Fla. Agric. Res. Educ. Cent. Res. Rep. BRA 1982-21.

Harris, D. C., and Davies, D. L. 1987. A disease of *Vinca rosea* caused by *Rhizopus stolonifer*. Plant Pathol. 36:608-609.

Ogawa, J. M., and English, H. Diseases of temperate zone tree and nut crops. Univ. Calif. Div. Agric. Nat. Resour. Publ. 3345.

Sarbhoy, A. K. 1966. *Rhizopus stolonifer*. Descriptions of Pathogenic Fungi and Bacteria, no. 110. Commonwealth Mycological Institute, Kew, England.

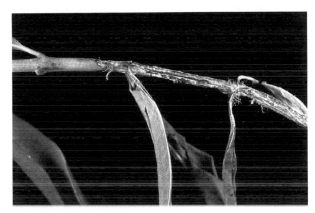

Fig. 11. Sporulation of *Rhizopus stolonifer* on a vinca (*Catharanthus roseus*) canker.

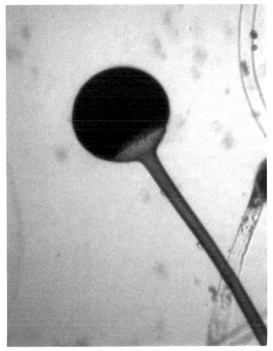

Fig. 12. Young sporangium of *Rhizopus stolonifer*.

Scab (Spot Anthracnose) of Poinsettia

Poinsettia scab (or spot anthracnose) was first reported on *Euphorbia pulcherrima* Willd. ex Klotzsch from Florida in 1941. This disease was a limiting factor in Florida poinsettia production prior to the increased availability of disease-free cuttings during the late 1960s.

Symptoms

Brown spots up to 4 mm in diameter with 1- to 2-mm chlorotic halos that buckle out from the leaf undersurface occur on poinsettia leaves affected by scab (Plate 49). Large areas of necrosis develop at leaf tips or margins, and the whole leaf may become chlorotic and abscise. Scablike, raised, circular or elongated (1–10 mm) lesions occur on stems, petioles, and leaf midribs (Plate 50). These lesions are tan and usually red or purple rimmed. Stem lesions may become sunken, develop fungal growth and sporulation, coalesce, and girdle stems. Occasionally, infected branches become abnormally elongated.

Causal Organism

Sphaceloma poinsettiae Jenk. & Ruehle has acervular conidiomata with a compact layer of pointed, pale to brown conidiophores arising from a fairly loose stroma. The conidiophores (15–30 × 3–5 μm) sometimes have one septum but are usually continuous. The minute conidia (7–20 × 2.5–5.3 μm) are oblong to elliptical (less often cylindrical or spherical); have one, two, or possibly three cells; are frequently constricted at the septa; and are either pale or as dark as the conidiophores.

Host Range and Epidemiology

S. poinsettiae is distributed throughout subtropical and tropical America and the South Pacific on the genus *Euphorbia*, occurring on wild species as well as cultivated ornamentals. In inoculation trials, *E. heterophylla* L. (Mexican fire plant), *E. prunifolia* Jacq. (painted euphorbia), *Manihot esculenta* Crantz (cassava), and *M. carthaginensis* Muell. were all susceptible to isolates of *S. poinsettiae* from *E. pulcherrima*.

Symptoms are visible 7 days after stem inoculation. The younger leaves are more susceptible than the older leaves. In one study, conidia were produced 1 month after inoculation. Conidia are spread from plant to plant by splashing water.

Management

Disease-free cuttings should be used, and they should not be carried over diseased stock plants. Mancozeb and thiophanate-methyl are registered in the United States for scab control on poinsettia. Objectionable residues from wettable powders may be reduced by adding a spreader-sticker.

Selected References

Engelhard, A. W. 1983. Control of poinsettia scab with fungicides. Fungic. Nematicide Tests 38:182.

Engelhard, A. W. 1984. Diseases and their control. Pages 37–47 in: Commercial Poinsettia Production in Florida. B. O. Tjia, ed. Univ. Fla. Dep. Ornamental Hortic. Publ. SP27.

Jenkins, A. E. 1942. Poinsettia scab discovered in Honolulu. Phytopathology 32:336–337.

Jenkins, A. E., and Ruehle, G. D. 1942. A new species of *Sphaceloma*—On poinsettia. Proc. Biol. Soc. Wash. 55:83–84.

Ruehle, G. D. 1941. Poinsettia scab caused by *Sphaceloma*. Phytopathology 31:947-948.

Stemphylium Leaf Spot of *Kalanchoe*

Stemphylium leaf spot of *Kalanchoe blossfeldiana* Poelln. was reported from Florida in 1962. The disease symptoms closely resemble those of physiological edema and thus may easily be misidentified.

Symptoms

Scablike, brown to black, raised lesions (1–3 mm) appear scattered across either leaf surface (Fig. 13). In some cases these spots enlarge, but they often remain small. Fungal sporulation on these areas is uncommon. Accurate diagnosis generally requires culturing. Conidiophores develop on fallen leaves, usually in clusters of three to 15, on the site of the original lesion.

Causal Organism

Stemphylium bolickii Sobers & C. P. Seymour has branched, septate, olivaceous hyphae; the septate conidiophores (98–392 × 3–6 μm) have bulbous tips. Nodular swellings are visible along the conidiophore. The conidia (30–56 × 13–21 μm) are usually olivaceous, oblong, and muriform with verrucose walls and one to three transverse septa. Ordinarily one septum (rarely two) is deeply constricted. The conidial apexes are obtuse or subacute and usually consist of one cell.

A closely related fungus, *S. floridanum* Hannon & G. F. Weber f. sp. *kalanchoe*, has been identified as a pathogen of *K. fedtschenkoi* Raym.-Hamet & E. Perrier (lavender-scallops) and *K. laxiflora* Bak. This species is distinguished from *S. bolickii* by the shape of the conidial apexes and the number of septa. Chlorosis and abscission are associated with infection by *S. f. kalanchoe*.

Host Range and Epidemiology

Natural infections by *S. bolickii* have been observed on *K. blossfeldiana* cultivars Tom Thumb, Red, and Yellow as well as other *Kalanchoe* species. *Echeveria*, *Sedum*, and *Crassula* spp. are also hosts of *S. bolickii*. Spores are produced on fallen leaves and may be spread by splashing during overhead irrigation. High temperature and high humidity favor infection.

Management

Foliage should be kept dry, plants should be sufficiently spaced, and infected leaf debris should be removed. Fungicides have not been tested for the control of this disease.

Fig. 13. Leaf spots on kalanchoe caused by *Stemphylium bolickii*. (Courtesy J. McRitchie)

TABLE 7. *Phyllosticta* spp. that Cause Leaf Spots on Flowering Potted Plants

Host	Species
Catharanthus roseus (vinca)	*Phyllosticta* sp.
Clerodendrum thomsoniae (bleeding heart vine)	*Phyllosticta* sp.
Cyclamen persicum	*P. cyclaminicola*
	P. cyclaminis
Dahlia hybrids	*P. dahliaecola*
Euphorbia pulcherrima (poinsettia)	*Phyllosticta* sp.
Eustoma grandiflorum (lisianthus)	*Phyllosticta* sp.
Gardenia augusta	*Phyllosticta* sp.
Gerbera jamesonii (African daisy)	*Phyllosticta* sp.
Hydrangea macrophylla	*P. hydrangeae*
Impatiens wallerana	*Phyllosticta* sp.
Pelargonium × *hortorum* (florist's geranium)	*Phyllosticta* sp.
Primula obconica (German primrose)	*P. primulicola*
Solanum pseudocapsicum (Jerusalem cherry)	*P. dulcamarae*
	P. pseudocapsici

Root and Crown Rot Diseases

Several soilborne fungi are responsible for economically important root and crown rot diseases of flowering potted plants (Table 9). The severity of these diseases is influenced by the virulence of the pathogen; physical and chemical properties of the growing medium; soil moisture, temperature, and pH; nutrition factors; and susceptibility of the host. Disease is most likely to occur when the soil environment is favorable for fungal germination and growth and suboptimal for the host. Soilless media can be particularly conducive to root rot pathogens because of reduced competition from nonpathogenic microorganisms. The same can be said for fumigated or overly heated field soils used in potting mixtures.

The fungi most frequently involved in root and crown rots of flowering potted plants include *Fusarium*, *Rhizoctonia*, *Sclerotinia*, and *Thielaviopsis*. The common root and crown diseases caused by *Pythium* and *Phytophthora* are discussed in the section on oomycete diseases.

Symptoms of root rot on aboveground plant parts are products of a debilitated root system and are usually expressed as nutrient deficiency symptoms, poor growth, chlorosis, wilt, or plant death. The symptoms are dependent on the host and the extent of root damage. Aboveground symptoms may in part be caused by plant growth regulator imbalances. Roots produce plant growth hormones, such as cytokinins and gibberellins, which, when carried to the aboveground part of the plant, influence stem elongation or the development and retention of leaves, buds, and flowers. When roots fail to develop adequately because of disease or poor soil conditions, the synthesis of these hormones may be reduced and symptoms characteristic of hormone deficiencies may occur.

Cylindrocarpon Root, Corm, and Petiole Rot of *Cyclamen*

Cylindrocarpon root, corm, and petiole rot of cyclamen (*Cyclamen persicum* Mill.) has been reported from Europe and the United States. This disease is generally less important than Fusarium wilt but periodically causes significant crop losses.

Symptoms

Infected cyclamen plants are stunted. Elliptical, dry, brown cankers approximately 1 cm long may form at the bases of petioles; corms and roots are also attacked. On the corms, shallow, depressed, dark brown cankers are formed. Roots may rot off next to the corm. Expansion of the petiole cankers and root invasion may lead to wilting of one or more leaves, so the disease may be mistaken for Fusarium wilt. Cracks may form on the top or sides of diseased corms. Growth of diseased plants is uneven.

Causal Organism

Cylindrocarpon destructans (Zinssmeister) Scholten (syns. *C. radicicola* Wollenweb. and *Ramularia destructans* Zinssmeister) is the imperfect state of *Nectria radicicola* Gerlach & L. Nilsson. It has curved, cylindrical, hyaline conidia similar to the macrospores of *Fusarium*, except that both ends of the conidia are rounded. *C. destructans* typically has conidia (20–60 × 4–6 μm) with three septa and also forms microconidia and chlamydospores. The optimum temperature for mycelial growth is 22°C.

Host Range and Epidemiology

C. destructans has a wide host range, including *Begonia*, *Campanula*, *Pelargonium*, *Primula*, *Saintpaulia* (Plate 56), and *Sinningia*. The fungus survives as a saprophyte and by producing chlamydospores in field soil and has been observed to most readily attack plants stressed by poor drainage, high levels of soluble salts, or phosphorous, potassium, or calcium deficiency.

Management

General sanitation practices are beneficial for avoiding introduction of inoculum. A well-drained potting medium should be used.

Selected References

Gerlach, W. 1956. Beitrage zur Kenntnis Gattrung Cylindrocarpon Wr. I. *Cylindrocarpon radicicola* Wr. als Krankheitserreger an Alpenveilchen. Phytopathol. Z. 26:161-170.

Rampanini, G. 1991. La coltivazione del ciclamino: Non tutto, ma quasi. Clamer Informa. Suppl. 9. Pentagono Editrice, Milan.

Scholten, G. 1964. *Nectria radicicola* en *Thielaviopsis basicola* als parasieten van *Cyclamen persicum*. Neth. J. Plant Pathol. Suppl. 2.

Fusarium Root, Crown, and Stem Rots

Fusarium spp. cause root, crown, and stem rots, and some are also capable of causing systemic vascular wilt disease (see Vascular Wilts Caused by *Fusarium oxysporum*). *Fusarium* is widely distributed in soils and is commonly isolated from roots and stems of many plants. *Fusarium* spp. are commonly encountered as plant pathogens, although many strains are strictly saprophytic, making diagnosis uncertain.

Symptoms

F. oxysporum, *F. solani*, *F. nivale*, and *F. roseum* cause cutting rot on florist's geranium, *Pelargonium* × *hortorum* L. H. Bailey. The rot occurs prior to root emergence and reportedly does not occur after roots develop. Lesions are found below the surface of the rooting medium and are dark black and soft. *F. oxysporum* has also been isolated from the roots of young seedling geraniums. The fungus causes a root and crown rot and eventually kills the plants.

F. oxysporum causes water-soaked lesions on the stems of holiday cactus (*Schlumbergera* and *Hatiora* spp.) at the soil line (Plate 57). As the disease progresses, orange to brown spots appear on the stems and leaves, which may be followed by blighting of the entire plant. Symptoms have also been

Selected References

McRitchie, J. J. 1984. Stemphylium leaf spot of Kalanchoe. Fla. Dep. Agric. Consumer Serv. Div. Plant Indus. Plant Pathol. Circ. 266.

Sobers, E. K. 1965. A form of *Stemphylium floridanum* pathogenic to species of *Kalanchoe*. Phytopathology 55:1313-1316.

Sobers, E. K., and Seymour, C. P. 1963. Stemphylium leaf spot of *Echeveria*, *Kalanchoe*, and *Sedum*. Phytopathology 53:1443-1446.

Stem Canker Diseases

Nectria Canker of *Exacum*

A stem rot and wilt of *Exacum affine* Balf. f. (exacum or Persian violet) caused by *Nectria haematococca* was first noted in California in 1980 and has since been observed in greenhouse crops of exacum in other parts of the United States and Canada. This disease, along with Botrytis blight and impatiens necrotic spot, has caused significant losses of exacum crops in the United States.

Symptoms

Nectria canker is evidenced by a brownish basal stem rot, often with fruiting bodies of the pathogen (asexual and/or sexual state). Pale pinkish sporodochia (Fig. 14) accompanied by dark red perithecia (Plate 51) assist in diagnosis of the disease, distinguishing it from cankers caused by *Botrytis cinerea* or tospovirus. The canker may progress down a branch into the main stem. Vascular discoloration may also be present. Exacum with Nectria canker have been observed to flower earlier than healthy plants (Plate 52). Lower stem rot and perithecia are also observed on *Gerbera jamesonii* H. Bolus ex Adlam (African daisy) with Nectria canker (Plate 53).

Causal Organism

N. haematococca Berk. & Broome (Fig. 15) is the perfect stage of *Fusarium solani* (Mart.) Sacc. (see Fusarium Root, Crown, and Stem Rots). Upon isolation, conidia develop within 7 days; orange perithecia form after 2–3 weeks on water agar with barley straw. Perithecia are superficial and ostiolate and appear to have a roughly warted outer wall and little to no visible stroma. Dry perithecia measure 130–200 (110–250) μm in diameter. The asci (60–80 × 8–12 μm) are cylindrical or clavate and unitunicate. The ascospores (11–18 × 4–7 μm) are hyaline to light brown, two celled, slightly constricted at the septum, ellipsoidal, and longitudinally striate.

F. solani has abundant, stout, thick-walled macroconidia, which are generally cylindrical so that the dorsal and ventral surfaces are parallel for most of their length. They have a blunt, rounded apical cell and a rounded, foot-shaped or notched basal cell and are borne on monophialides. Microconidia (8–16 × 2–4 μm) are primarily one celled, oval to kidney shaped, and larger with thicker walls than microconidia of *F. oxysporum*. They may be few or abundant, depending upon the isolate. The microconidia of *F. solani* are borne on long monophialides (45–80 × 2.5–3 μm) on elongate, sparsely branched microconidiophores (up to 400 μm). The microconidiophores distinguish this species from *F. oxysporum*, which has short microconidiophores terminating in multiple phialides. Single and paired chlamydospores (9–12 × 8–10 μm) are usually abundant in *F. solani* and are terminal or intercalary.

F. solani grows relatively fast on potato-dextrose agar and often produces profuse aerial mycelium (Plate 54). The surface of the colony turns cream, blue green, or blue (but not orange or purple) after the production of confluent sporodochia. The colony undersurface is usually colorless or may be dark violet.

Host Range and Epidemiology

F. solani is a pathogen with a wide host range, although some strain specificity is known. *N. haematococca*, the teleomorph of the *F. solani*, is less frequently encountered as a plant pathogen, and the host range of the exacum isolate remains undetermined. An attempt was made to inoculate the chrysanthemum (*Dendranthema* × *grandiflorum* Kitam.) cultivar Butterball with an exacum isolate, but no symptoms developed.

On exacum, disease is reported to be more severe at 32°C than at 21 or 27°C. Wounding does not appear to be necessary for infection, but observations in commercial greenhouses indicate that pruning wounds are associated with serious disease outbreaks. Root dips into inoculum prior to transplanting resulted in wilt without visible root decay. Symptomless inoculated plants developed symptoms and died after being moved from 21 to 32°C.

Fig. 14. Sporodochia of the *Fusarium solani* stage of *Nectria haematococca* on an exacum canker.

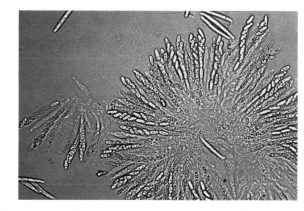

Fig. 15. Asci of *Nectria haematococca*. (Courtesy A. W. Engelhard)

Management

Wounding of exacum during production should be avoided, and wilted individuals should be rogued to reduce inoculum. Fungus gnats or other insects that may carry spores and provide wounds for fungus infection should be controlled.

Selected References

Booth, C. 1971. The Genus *Fusarium*. Commonwealth Mycological Institute, Kew, England.

Nelson, P. E., Toussoun, T. A., and Cook, J. J., eds. 1981. *Fusarium: Diseases, Biology, and Taxonomy.* The Pennsylvania State University Press, University Park.

Nelson, P. E., Toussoun, T. A., and Marasas, W. F. O. 1983. *Fusarium* Species: An Illustrated Manual for Identification. The Pennsylvania State University Press, University Park.

Pierce, L., and McCain, A. H. 1982. Stem rot and wilt of *Exacum affine* caused by *Nectria haematococca*. Plant Dis. 66:161-163.

Phomopsis Canker of *Gardenia*

Phomopsis canker of *Gardenia augusta* (L.) Merrill was first reported in the 1930s from California and Ohio. It was considered the most common disease of gardenias during the 1940s, when roughly a million square feet of greenhouse space was devoted to cut-flower and pot-plant production of gardenias in New Jersey alone. The disease has also been reported from Europe.

Symptoms

Sunken, brown, dead areas appear at the base of the stem. As cankers develop, the affected tissue becomes swollen and corky, and the bark splits (Fig. 16 and Plate 55). The normally greenish white stem tissue beneath the bark develops an orange discoloration in the area of the canker. Cankers may also develop on branches. Foliage above the cankers wilts and

Fig. 16. Swollen stem base on gardenia caused by Phomopsis canker. (Courtesy M. Shurtleff)

dies once the cankers completely girdle the stem or branch. Black pycnidia are produced on the cankers and on leaf tissue contacting moist soil.

Causal Organism

Phomopsis gardeniae Buddin & Wakef., the anamorph of *Diaporthe gardeniae,* has nonostiolate, thick-walled pycnidia (300 μm in diameter) containing aseptate, hyaline conidiophores (15–20 × 4 μm) that taper to a pointed tip. Under moist conditions, hyaline, fusiform alpha-spores (7–12 × 2.5–4.0 μm) are exuded from the pycnidia in pale salmon or buff tendrils or masses. Beta-spores (25 × 1 μm) are also produced less frequently.

Host Range and Epidemiology

P. gardeniae is reported only from *G. augusta.* The fungus is a wound parasite and is not known to penetrate uninjured tissues. Infection typically begins during propagation at lower nodes from which the leaves have been stripped before planting. When these freshly wounded areas are buried in the rooting medium, they provide points of entry for the fungus. Reuse of contaminated rooting medium contributes to the disease. Rooted cuttings may show only a faint yellow discoloration at *Phomopsis*-infected nodes, thus allowing for infected plant material to be distributed.

Management

Gardenia cuttings should be rooted in a soilless medium or a soil-based medium that has been heat or chemically treated. Medium should not be reused unless it has been treated to eliminate pests. Wetting the foliage during irrigation and prolonged leaf wetness should be avoided. Plant spacing should allow good air circulation. Affected plants should be discarded. Mealybugs and scale insects should be controlled because they can be a source of wounds that facilitate entry of the fungus.

Selected References

Buddin, W., and Wakefield, E. M. 1937. A stem-canker disease of gardenias. Gard. Chron. 101:226-227.

Pirone, P. P. 1940. Diseases of the gardenia. N.J. Agric. Exp. Stn. Bull. 679.

Uecker, F. A. 1988. A World List of Phomopsis Names with Notes on Nomenclature, Morphology and Biology. Mycologia Memoir, no. 13. J. Cramer, Berlin.

Minor Leaf, Flower, and Stem Diseases

The leaves, stems, and flowers of potted plants are susceptible to many minor pathogens that have not been studied in detail. These are listed in Table 6 (*Stemphylium* spp.), Table 7 (*Phyllosticta* spp.), and Table 8 (miscellaneous fungi).

TABLE 6. *Stemphylium* spp. that Cause Leaf Spots on Flowering Potted Plants

Host	Species
Aquilegia caerulea (columbine)	*Stemphylium* sp.
Dicentra spectabilis (bleeding heart)	*Stemphylium* sp.
	S. lycopersici
Kalanchoe blossfeldiana	*S. bolickii*
Solanum pseudocapsicum (Jerusalem cherry)	*S. solani*

described as initially reddish lesions at the soil line that become papery and tan. Terminal branches may abscise. Both cuttings and established plants can become diseased.

Crown rot of African daisy (*Gerbera jamesonii* H. Bolus ex Adlam) caused by *F. oxysporum* has been described in Holland, Poland, and the United States. The stems of affected plants darken, turn black, and decay. Stems and leaves wither and wilt, leading to plant death. In Poland, the causal organism was reported as *F. oxysporum* f. sp. *gerberae.*

Root rots caused by *Fusarium* spp. are occasionally

TABLE 8. Miscellaneous Fungal Leaf, Stem, and Flower Diseases on Flowering Potted Plants

Host	Pathogen	Disease
Aquilegia caerulea (columbine)	*Haplobasidion thalictri*	Leaf spot
	Urocystis sorosporioides	Leaf smut
Bougainvillea spectabilis	*Cercosporidium bougainvilleae*	Leaf spot
	Nectria cinnabarina	Dieback
Clerodendrum thomsoniae	*Corynespora* sp.	Leaf spot
Crossandra infundibuliformis	*Helminthosporium* sp.	Flower spot
Cyclamen persicum	*Phoma exigua*	Damping-off
	Phyllosticta cyclaminicola	Leaf spot
	Phyllosticta cyclaminis	Leaf spot
Dahlia hybrids	*Ascochyta dahliicola*	Leaf spot
	Itersonilia perplexans	Flower blight
Euphorbia pulcherrima (poinsettia)	*Ascochyta* sp.	Leaf spot
	Fusarium solani	Stem rot
	Helminthosporium sp.	Leaf spot
	Lasiodiplodia theobromae	Twig dieback
	Nectriella pironii	Stem gall
	Phomopsis sp.	Stem canker
	Syspastospora parasitica	Leaf and stem rot
Eustoma grandiflorum (lisianthus)	*Fusarium solani*	Stem rot
	Sclerophoma eustomonis	Stem blight
Gardenia augusta	*Ascochyta doronici*	Leaf spot
	Erythricium salmonicolor	Limb blight
	Fusarium lateritium	Bud rot
	Fusarium roseum	Bud rot
	Diplodia sp.	Tip dieback
	Leptothyrium sp.	Tip necrosis
	Phoma sp.	Stem canker
	Sphaeropsis sp.	Leaf spot
Gerbera jamesonii (African daisy)	*Ascochyta doronici*	Leaf spot
	Fusarium solani	Stem rot
	Septoria sp.	Leaf spot
Hibiscus rosa-sinensis	*Botryosphaeria ribis*	Dieback
	Botryospheria sp.	Leaf spot
	Coniothyrium sp.	Dieback
	Curvularia sp.	Leaf spot
	Leptosphaeria agnita	Leaf spot
	Macrophoma sp.	Leaf spot
	Phomopsis sp.	Limb dieback, leaf spot
	Phyllostictina hibiscina	Leaf spot
	Physalospora sp.	Leaf spot
	Pseudocercospora abelmoschi	Leaf spot
	Sphaeropsis sp.	Witches'-broom
	Volutella sp.	Canker
Hydrangea macrophylla	*Helminthosporium* sp.	Leaf spot
	Nectriella pironii	Stem gall
	Phoma exigua	Leaf spot
	Phomopsis sp.	Canker
Lantana camara	*Mycovellosiella lantanae*	Leaf spot
Pelargonium peltatum (ivy geranium)	*Fusarium* sp.	Stem rot
Pelargonium × *hortorum* (florist's geranium)	*Bipolaris maydis*	Leaf and petal spot
	Bipolaris setariae	Leaf and petal spot
	Cylindrocladium scoparium	Flower and leaf spot
	Fusarium sp.	Stem rot
Pericallis × *hybrida* (cineraria)	*Ascochyta cinerariae*	Leaf spot
Schizanthus × *wisetonensis* (butterfly flower)	*Ascochyta* sp.	Leaf spot
Schlumbergera truncata (Thanksgiving cactus)	*Dichotomophthora* sp.	Tip blight
	Fusarium moniliforme	Basal stem rot, leaf spot
	Fusarium oxysporum	Dieback, root and stem rot
	Phomopsis sp.	Leaf spot
Sinningia speciosa (gloxinia)	*Cladosporium herbarum*	Leaf rot
	Plenodomus sp.	Leaf spot

encountered on other flowering potted plants, including exacum (*Exacum affine* Balf. f.), lisianthus (*Eustoma grandiflorum* (Raf.) Shinn.), and poinsettia (*Euphorbia pulcherrima* Willd. ex Klotzsch). Roots are softened and exhibit a brown to black discoloration (Plate 58). Stems infected by *F. solani* may show cream to pale orange sporodochia of the imperfect state and perithecia of the perfect state of the pathogen (Plate 59).

Causal Organisms

F. solani (Mart.) Sacc. is the most common cause of root, crown, and lower stem rots; however, forms of *F. oxysporum* Schlectend.:Fr. and other species of the genus can cause similar rots in flowering potted plants. The genus *Fusarium* is included in the Tuberculariaceae family of the fungi imperfecti. Descriptions of *F. solani* and *F. oxysporum* can be found in other sections of this book.

Host Range and Epidemiology

Various *Fusarium* species have been reported from flowering potted plants. Of these, only Fusarium rots on *P. × hortorum*, *G. jamesonii*, *E. affine*, and *Schlumbergera truncata* (Haw.) Moran (holiday cactus) have been described significantly in the literature. (See Nectria Canker of *Exacum* for a discussion of *F. solani* on that host).

In the absence of a living host, *Fusarium* persists saprophytically, as chlamydospores, or as resistant hyphae or

TABLE 9. Fungi that Cause Root and Crown Rots of Flowering Potted Plants

Host	Fungus	Host	Fungus
Anemone coronaria	*Rhizoctonia* sp. *Sclerotinia sclerotiorum* *Sclerotium rolfsii*	*G. jamesonii (continued)*	*R. solani* *Sclerotinia sclerotiorum* *Sclerotium rolfsii* *T. basicola*
Aquilegia spp. (columbine)	*R. solani* *Sclerotinia sclerotiorum* *Sclerotium rolfsii*	*Hibiscus rosa-sinensis*	*R. solani* *Sclerotinia sclerotiorum* *Sclerotium rolfsii*
Begonia sp.	*Fusarium* sp. *R. solani* *Sclerotinia sclerotiorum* *Sclerotium rolfsii* *Thielaviopsis basicola*	*Hydrangea macrophylla*	*R. solani*
		Impatiens sp.	*Fusarium* sp. *R. solani* *Sclerotium rolfsii*
Bougainvillea spp.	*Fusarium* sp. *R. solani*	*Kalanchoe blossfeldiana*	*Cylindrocladium* sp. *R. solani*
Calceolaria Herbeohybrida group (pocketbook plant)	*Sclerotinia sclerotiorum* *Myrothecium roridum*	*Lantana camara*	*R. solani*
Campanula carpatica (bellflower)	*Sclerotium rolfsii*	*Pelargonium × hortorum* (florist's geranium)	*F. nivale* *F. oxysporum* *F. roseum* *F. solani* *R. solani* *Sclerotinia sclerotiorum* *Sclerotium* sp. *T. basicola*
Capsicum annuum (ornamental pepper)	*R. solani* *Sclerotinia sclerotiorum* *Sclerotium rolfsii*		
Catharanthus roseus (vinca)	*R. solani* *Fusarium* sp.		
Crossandra infundibuliformis	*R. solani*	*Pericallis × hybrida* (cineraria)	*Fusarium* sp. *R. solani* *T. basicola*
Cyclamen persicum	*Cylindrocarpon destructans* *Cylindrocladiella peruviana* *F. oxysporum* *T. basicola*	*Primula* sp. (primrose)	*R. solani*
		Ranunculus sp.	*R. solani*
Dahlia hybrids	*Colletotrichum coccodes* *Fusarium* sp. *R. solani* *Sclerotinia sclerotiorum* *Sclerotium rolfsii*	*Saintpaulia ionantha* (African violet)	*Cylindrocarpon destructans* *F. solani* *Fusarium* sp. *R. solani*
		Schizanthus sp. (butterfly flower)	*R. solani*
Euphorbia pulcherrima (poinsettia)	*Cylindrocladium scoparium* *F. moniliforme* *F. solani* *Fusarium* sp. *T. basicola* *R. solani*	*Schlumbergera truncata* (Thanksgiving cactus)	*F. moniliforme* *F. oxysporum* *Fusarium* sp. *R. solani* *Rhizoctonia* sp.
Eustoma grandiflorum (lisianthus)	*R. solani* *F. solani*	*Sinningia speciosa* (gloxinia)	*M. roridum* *R. solani* *Rhizoctonia* sp. *Sclerotinia sclerotiorum* *Sclerotium rolfsii*
Exacum affine (Persian violet)	*F. solani* *R. solani* *Sclerotium rolfsii*		
Fuchsia × hybrida	*Rhizoctonia* sp. *T. basicola*	*Solanum pseudocapsicum* (Jerusalem cherry)	*R. solani* *Sclerotium rolfsii*
Gardenia augusta	*Cylindrocladium scoparium* *F. oxysporum* *R. solani*	*Verbena × hybrida*	*T. basicola* *R. solani* *Rhizoctonia* sp.
Gerbera jamesonii (African daisy)	*F. oxysporum* *F. solani*		

spores. Chlamydospores typically form within or on the surface of infected plant tissues. Macroconidia washed into the soil also transform into chlamydospores. Germination of chlamydospores is stimulated by exudates from plant roots and germinating seeds. Chlamydospores and spores germinate over a wide range of pH, but optimum temperature is 25–28°C. *Fusarium* will survive longer in dry soils than in wet soils. Spores are disseminated by air movement, insects, transportation of diseased plants, and splashing irrigation water.

In inoculation experiments with unrooted, partially rooted, and fully rooted cuttings of 22 geranium cultivars, *F. oxysporum*, *F. solani*, *F. nivale*, and *F. roseum* isolates were most virulent in combination with *Rhizoctonia solani* isolates. The *Fusarium* isolates alone caused decay of unrooted cuttings but not of rooted cuttings.

Management

Control of *Fusarium* diseases is best accomplished by preventing the fungus from contaminating the plant-production area. *Fusarium* usually enters the greenhouse via contaminated media, soiled containers, and plant material. It is probable that insects such as fungus gnats and shore flies introduce *Fusarium* into the growing medium. Sanitation practices should be initiated at propagation time and continued through the crop cycle. Soiled hands and tools should not come in contact with soilless media. Plant pathogens such as *Fusarium* develop rapidly when introduced into soilless media or rooting cubes because of the lack of competing microorganisms. Pasteurized field soil is also subject to rapid colonization by soilborne plant pathogens. In general, cleanliness of the plant-growing area is recommended. Diseased plants and plant debris should be removed as they appear on the bench; space between plants should be increased to improve aeration; and excessive splashing during watering should be avoided. General sanitation is especially critical with *Fusarium* because it is more easily disseminated through the air than *Rhizoctonia*. Chlorothalonil and thiophanate-methyl have been shown to provide control of Fusarium root and crown rots. Fungicide labels should be checked for approved uses.

Selected References

Booth, C. 1971. The Genus *Fusarium*. Eastern Press, London.

Manning, W. J., Vardaro, P. M., and Cox, E. A. 1973. Root and stem rot of geranium cuttings caused by *Rhizoctonia* and *Fusarium*. Plant Dis. Rep. 57:177-179.

Moorman, G. W., and Klemmer, R. A. 1980. *Fusarium oxysporum* causes basal stem rot of *Zygocactus truncatus*. Plant Dis. 64:1118-1119.

Nelson, P. E., Toussoun, T. A., and Cook, R. J., eds. 1981. *Fusarium: Diseases, Biology, and Taxonomy*. The Pennsylvania State University Press, University Park.

Nelson, P. E., Toussoun, T. A., and Marasas, W. F. O. 1983. *Fusarium* Species, An Illustrated Manual for Identification. The Pennsylvania State University Press, University Park.

Raabe, R. D., Hurlimann, J. H., and Bruckner, B. 1975. A root and stem rot of Christmas cactus and its control. Pages 1-2 in: Calif. Plant Pathol. 38.

Diseases Caused by *Rhizoctonia solani*

Rhizoctonia solani is widely distributed, has a wide host range, and commonly causes damping-off, root rot, crown rot, and leaf and stem blight. When the fungus grows over the foliage of the plant, the disease is referred to as web blight (Plate 60). *Rhizoctonia* occurs on most flowering potted plants and is one of the most common plant pathogens encountered.

Symptoms

When *R. solani* causes damping-off, stems become darkened at the soil line and may become shrunken to a wire-stem appearance (Fig. 17). Under warm, humid conditions, the fungus will grow over the entire seedling, sometimes encompassing large portions of seedling trays (Plates 61 and 62). When this occurs, a light brown mycelium, closely appressed to the plant tissues and soil, is usually apparent. Since the fungus grows out radially from a point source of contamination, damage is usually in a circular or arc-shaped pattern. *Catharanthus roseus* (L.) G. Don (vinca), *Begonia*, and *Impatiens* spp. are particularly susceptible to damping-off.

R. solani may also cause a web blight on small seedlings or larger plants that have a dense canopy in the absence of damping-off or root rot. The mycelium appears as a brown, cobweb growth on the leaves; after infection, the leaves turn brown and rot (Plates 60, 63, and 64). When leaves are in contact with the growing medium or soil is splashed onto the foliage, a leaf spot may occur (Plate 65).

Root rots caused by *R. solani* are characterized by discrete, brown lesions and rotting of the cortical tissues. New root development is inhibited (Plate 66). When root rot is extensive, discrete lesions may no longer be evident. *R. solani* causes crown rot and cutting rot of potted plants more often than root rot, although poinsettia is quite susceptible to Rhizoctonia root rot.

Potted plants propagated as stem cuttings, such as geranium, poinsettia, and New Guinea impatiens, are susceptible to Rhizoctonia cutting rot. Disease is evident as a dry, brown basal rot that may develop before or after rooting. On geranium, the brown coloration helps to distinguish this disease from the more common Pythium black leg, which generally results in a black canker at the stem base. African violets are also susceptible to infection by *R. solani* during vegetative propagation (Plate 67).

Crown rot can result from progression of the pathogen from the roots into the crown or from direct fungal penetration of the crown tissues. *R. solani* commonly causes crown rot in the absence of root rot. On *Euphorbia pulcherrima* Wild. ex Klotzsch (poinsettia), *Begonia*, *Saintpaulia ionantha* Wendl. (African violet), impatiens, New Guinea impatiens, and many other hosts, *R. solani* causes a brown crown canker. Longitudinal cracking and a dry appearance of the rotted crown tissues often develops on older plants (Plate 68). Other aboveground symptoms include chlorosis, wilting, loss of lower leaves, and at times stunting of the entire plant. Plant death commonly occurs. Crown rot often develops after plants have been transplanted into the landscape.

Causal Organism

Definitive morphological characteristics of *R. solani* Kühn (teleomorph *Thanatephorus cucumeris* (A. B. Frank) Donk)

Fig. 17. Wire-stem appearance of vinca caused by *Rhizoctonia solani*.

include the presence of multinucleate cells in young vegetative hyphae, branching near the distal septum, constriction of the branch, formation of a septum near the point of branch origin, and mycelium with some shade of brown (Fig. 18). Cultures on potato-dextrose agar show a fleecy margin, vary in color from light tan to medium brown, and may develop sclerotia (Plate 69). The following characteristics are usually present, but one or more may be lacking: monilioid cells, sclerotia without a differentiated rind and medulla, hyphae greater than 5 μm in diameter, rapid growth rate, and pathogenicity to plants. *Rhizoctonia*-like organisms with clamp connections, conidia, differentiated sclerotia, rhizomorphs, pigmentation other than brown, and teleomorph stages other than *T. cucumeris* are not included in the *R. solani* group. *T. cucumeris*, family Corticeaceae of Basidiomycotina, is not commonly encountered.

Anastomosis grouping has provided an indication of relationships within groups of isolates, allowing separation of *R. solani* into genetically independent entities. Most isolates from herbaceous ornamentals occur in anastomosis groups AG4 and AG1. The reported isolates of *R. solani* from New Guinea impatiens all belong to AG4.

Host Range and Epidemiology

Rhizoctonia root rot occurs on most flowering potted plants and causes considerable economic losses. Root rot, crown rot, and/or web blight are commonly encountered on *Begonia* sp., *C. roseus*, *Gerbera jamesonii* H. Bolus ex Adlam (African daisy), *S. ionantha*, *Campanula carpatica* Jacq. (beliflower), *E. pulcherrima*, *Impatiens wallerana* Hook f. (impatiens), New Guinea hybrid impatiens, and *Pelargonium × hortorum* L. H. Bailey (florist's geranium). A list of flowering potted plants susceptible to Rhizoctonia root, crown, and basal stem rot is provided in Table 9.

Disease incidence and severity is dependent on the host, fungal strain, and environmental conditions. Poinsettias are most susceptible just before or soon after rooting and just prior to plant maturity. *R. solani* generally grows best in soils that are evenly moist and warm. In studies of environmental influences on disease development, Rhizoctonia root rot increased at soil temperatures of 17–26°C and when the moisture-holding capacity of the soil was below 40%. *R. solani* also grows best in soils with high oxygen and low carbon dioxide levels.

Management

Sanitation practices outlined for *Fusarium* are also applicable to *Rhizoctonia*. Fungicides such as iprodione, pentachloronitrobenzene, thiophanate-methyl, and triflumizole are effective in drench treatments against *Rhizoctonia*. Product labels should be checked for approved uses. When soil applications of fungicides to control root and crown rots are used, treatments must be made before infection occurs. Chlorothalonil is registered in the United States for spray application

to control Rhizoctonia web blight on some flowering potted plants.

A difference in susceptibility to *R. solani* infection among New Guinea impatiens cultivars has been observed. In an inoculation experiment with the cultivars Astro, Aurora, Columbia, Corona, Cosmos, Equinox, Gemini, Milky Way, Nova, Red Planet, Sunset, Twilight, and Twinkle, Milky Way and Gemini grew best and were more resistant to the disease than the others tested. Cultivars Astro, Aurora, Nova, and Sunset were most susceptible. In general, however, resistance is not a practical means of controlling *R. solani* in potted plants.

Selected References

Bateman, D. F. 1961. The effect of soil moisture upon development of poinsettia root rots. Phytopathology 51:445-451.

Bateman, D. F. 1961. Environment and the poinsettia root rots. Pages 1-4 in: N.Y. State Flower Grow. Bull. 186.

Bateman, D. F. 1962. Relation of soil pH to development of poinsettia root rots. Phytopathology 52:559-566.

Benson, D. M. 1991. Control of Rhizoctonia stem rot of poinsettia during propagation with fungicides that prevent colonization of rooting cubes by *Rhizoctonia solani*. Plant Dis. 75:394-398.

Castillo, S., and Peterson, J. L. 1990. Cause and control of crown rot of New Guinea impatiens. Plant Dis. 74:77-79.

Cavileer, T. D., and Peterson, J. L. 1988. Isolation and characterization of *Rhizoctonia solani* and binucleate *R. solani*-like fungi from herbaceous and woody ornamentals in New Jersey. (Abstr.) Phytopathology 78:1506.

Parmeter, J. R., Jr., ed. 1970. *Rhizoctonia solani*, Biology and Pathology. University of California Press, Berkeley.

Powell, C. C., Jr. 1988. The safety and efficacy of fungicides for use in Rhizoctonia crown rot control of directly potted unrooted poinsettia cuttings. Plant Dis. 72:693-695.

Sneh, B., Burpee, L., and Ogoshi, A. 1991. Identification of *Rhizoctonia* Species. American Phytopathological Society, St. Paul, MN.

Diseases Caused by *Sclerotinia sclerotiorum*

Sclerotinia sclerotiorum infects more than 360 species of plants, among them several important vegetable and flower crops as well as weeds. *S. sclerotiorum* is a problem mostly in field soil outdoors but may cause significant root, crown, and stem rots of a variety of plants growing in the greenhouse. The fungus also causes preemergence and postemergence damping-off of seedlings.

Symptoms

S. sclerotiorum produces conspicuous white strands of mycelium that occasionally form a dense mat on the soil surface or at the base of a stem (Plate 70). Infection of the foliage may occur when leaves are in contact with the ground.

S. sclerotiorum causes a crown rot of florist's geranium (*Pelargonium × hortorum* L. H. Bailey). Disease is evidenced by a cottony covering of fungal hyphal growth on the soil that continues up the stem. Underlying tissues are decayed, and black sclerotia soon develop. The disease also occurs on seedling geraniums.

Causal Organism

S. sclerotiorum (Lib.) de Bary is an ascomycete in the order Helotiales. It does not have an anamorphic state. The teleomorph consists of a small, cup-shaped, brown apothecium 2–4 mm in diameter on a stalk 2–2.5 cm in length (Fig. 19). The apothecia arise from a sclerotium that consists of a thin, dark brown to black rind comprised of three to four layers of thick-

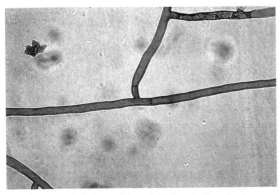

Fig. 18. Characteristic branching of *Rhizoctonia solani*.

walled cells encompassing a stromalike mass of white, pseudoparenchymatous, thin-walled cells. The large sclerotia form readily in culture on potato-dextrose agar (Fig. 20).

Host Range and Epidemiology

Potted plants reported to be hosts of *S. sclerotiorum* are listed in Table 9. It is likely that this list does not reflect the potential host range of flowering potted plants, since this fungus is known to affect many vegetable and flower genera.

The primary source of infection is sclerotia, which exist in soil and plant soil debris. The sclerotia vary in size but are commonly 5–30 × 3–10 mm and are quite resistant to extremes in the environment such as dryness and cold. Optimum temperature for germination of sclerotia is 13–15°C, but a few will germinate at 4–26°C. Germination is also dependent on high soil moisture level for an extended period of time. It has been reported that in outdoor growing areas, soil must be at field capacity for 10 days before germination will occur.

Sclerotia have two modes of infection. They may infect directly, or they may produce fruiting structures (apothecia) that disseminate ascospores. Apothecia do not typically form under greenhouse conditions.

Management

When field soil is a component of the growing medium, it must be steam treated or fumigated. Thiophanate-methyl is registered in the United States for drench treatments to control *Sclerotinia* on flowering potted plants.

Selected References

Engelhard, A. W. 1993. Cottony stem rot. Page 255 in: Geraniums IV. J. W. White, ed. Ball Publishing, Geneva, IL.

Kohn, L. M. 1979. A monographic revision of the genus *Sclerotinia*. Mycotaxon 9:365-444.

Orlikowski, I. 1976-1977. Investigations of causal agents of African daisy diseases in greenhouses in Poland. (In Polish, with English summary.) Pr. Inst. Sadow. Skierniewicach Ser. B 2:197-201.

Diseases Caused by *Thielaviopsis basicola*

Thielaviopsis basicola is a widely distributed root pathogen reported from at least 12 plant families but most frequently associated with Fabaceae, Solanaceae, and Cucurbitaceae. Many ornamental plants are susceptible to Thielaviopsis root rot. In the potted-plant industry, *T. basicola* is more commonly encountered when the growing medium has field soil as a component.

Symptoms

Symptoms aboveground are not unique to *T. basicola* but are common to other causes of root disease or unfavorable growing conditions. Depending on the susceptibility of the host and the degree of root rot, wilting may or may not occur (Plates 71 and 72). Plants are often chlorotic and stunted (Plate 73). Roots severely affected by *T. basicola* are darkly discolored, hence the common name, black root rot (Plates 74 and 75; Fig. 21). Root symptoms can be diagnostic. When abundant chlamydospores and pseudoparenchymatous stromatic tissues develop on the surface of the root lesion, a distinct blackening occurs. However, when the darkly pigmented structures are not formed in abundance or when they are formed deeper in the root cortex, lesions are brown and may be difficult to distinguish from those caused by *Rhizoctonia*.

On *Cyclamen persicum* Mill., *T. basicola* causes black spotting and discoloration of roots. The small feeder roots are most commonly infected. Severely affected plants grow slowly, but the fungus generally does not invade cyclamen corms or leaf petioles.

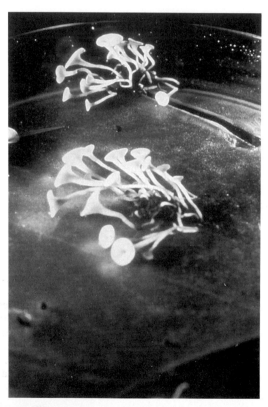

Fig. 19. Apothecia formed on sclerotia of *Sclerotinia sclerotiorum* in culture. (Courtesy M. Gleason)

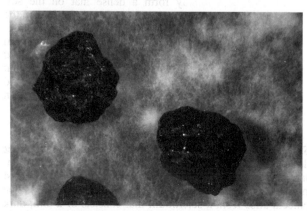

Fig. 20. Black sclerotia of *Sclerotinia sclerotiorum.*

Fig. 21. Blackening of a secondary root caused by *Thielaviopsis basicola.*

T. basicola causes severe root rot in African daisy (*Gerbera jamesonii* H. Bolus ex Adlam). The disease has been reported from Europe and the United States. Leaves of affected plants are stunted and distorted and become yellow around the margins. Infected roots are at first spotted and become black with age.

Although it has been widely reported that poinsettias (*Euphorbia pulcherrima* Willd. ex Klotzsch) are more susceptible as they approach maturity, plants are susceptible to colonization at all stages of growth. Thielaviopsis root rot of poinsettia results in stunted, chlorotic plants that may defoliate. Roots have brown to black lesions (Plate 75). Longitudinal cracking of the stem originating beneath the soil level and extending above is characteristic of this disease on poinsettia. The lower stem may also be blackened because of the accumulation of chlamydospores (Plate 76). Leaf spots may develop during periods of high humidity.

Causal Organism

T. basicola (Berk. & Broome) Ferraris (syn. *Chalara elegans* Nag Raj & Kendrick) is a dematiaceous hyphomycete readily identified by its cylindrical endoconidia (Fig. 22) and dark brown chains of chlamydospores (Fig. 23 and Plate 77). Conidia (7.5–19 × 3–5 μm) are borne within the conidiophore and extruded in long chains. They are unicellular, hyaline, and cylindrical to doliform with truncate or obtuse ends. Chlamydospores (6.5–14 × 9–13 μm) are arranged in a linear series of five to seven spores. They are unicellular and dark brown, have a thin outer and thicker inner wall (1–2 μm), and possess transverse germ pores at each end of the individual spore. Chlamydospores are short cylindrical, except for the apical chlamydospore, which is conoid.

For prompt diagnosis, the fungus can be observed within infected tissue with the aid of a compound microscope. Direct isolation of the fungus on culture media is sometimes difficult. *T. basicola* can be readily isolated by spreading a thin layer of soil or colonized root tissue over 5-mm-thick carrot root disks in sterile petri dishes. Sterile water is then atomized over the disks to moisten the soil or roots, and after 2–4 days at room temperature, the incubated disks are rinsed with water and incubated again. If the fungus is present, endoconidia are formed in approximately 6 days followed by the formation of chlamydospores. Beadlike chains of endoconidia can be observed with a dissecting microscope. Isolates on potato-dextrose agar are slow growing and usually develop a brown pigment, but sectoring to a less pigmented mycelium has been observed in cultures maintained over time (Plate 78).

Host Range and Epidemiology

Flowering potted plant hosts of *T. basicola* are listed in Table 9. Diseases of *C. persicum*, *G. jamesonii*, and *E. pulcherrima* are most commonly referred to in the literature.

T. basicola forms chlamydospores, which allow survival in soil. Chlamydospores are stimulated to germinate by root exudates. Germination will occur over a range of soil pH from 5.0 to 8.5.

Root rot is most significant at soil temperatures of 13–17°C, but some researchers have reported optimum temperatures as high as 25°C. Moderate disease development may occur when the moisture-holding capacity of the soil is at 36%, but the disease is significantly more serious at 70% and above. Root rot is also more serious in plants grown in neutral and alkaline soils. The disease is greatly reduced at pH levels below 5.5.

Untreated or improperly treated field soil is probably the primary source of *T. basicola* in potted-plant culture. However, disease can develop in soilless media, and there are reports of *T. basicola* contamination of commercial peat moss.

Management

Sanitation principles outlined for *Rhizoctonia* are also applicable to *T. basicola*. Fungus gnats and shore flies may serve as vectors. When field soil is a component of the growing medium, it should be treated with steam or chemical fumigants. Thiophanate-methyl and triflumizole fungicides are registered for use in the United States against *Thielaviopsis* on some flowering potted plants.

Variation in susceptibility to *T. basicola* has been reported; but in general, resistance is not a practical method of control. Red-flowered cyclamen are generally more resistant to *T. basicola* than white-flowered types.

Selected References

Bateman, D. F. 1961. The effect of soil moisture upon development of poinsettia root rots. Phytopathology 51:445-451.

Bateman, D. F. 1961. Environment and the poinsettia root rots. Pages 1-4 in: N.Y. State Flower Grow. Bull. 186.

Bateman, D. F. 1962. Relation of soil pH to development of poinsettia root rots. Phytopathology 52:559-566.

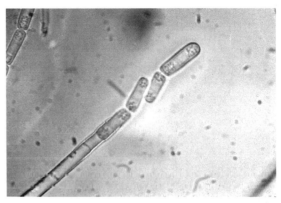

Fig. 22. Endogenously formed conidia of *Thielaviopsis basicola*.

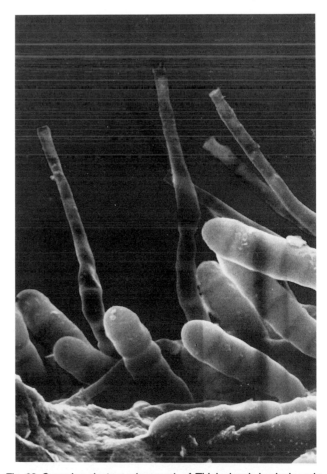

Fig. 23. Scanning electron micrograph of *Thielaviopsis basicola* endoconidiophores and chlamydospores on a root surface.

Bateman, D. F., and Dimock, A. W. 1959. The influence of temperature on root rots of poinsettia caused by *Thielaviopsis basicola*. Phytopathology 49:641-647.

Keller, J. R., and Potter, H. S. 1954. *Thielaviopsis* associated with root rots of some ornamental plants. Plant Dis. Rep. 38:354-355.

Linderman, R. G. 1993. Thielaviopsis black root rot. Pages 255-262 in: Geraniums IV. J. W. White, ed. Ball Publishing, Geneva, IL.

Mathre, D. E., and Ravenscroth, A. V. 1966. Physiology of germination of chlamydospores and endoconidia of *Thielaviopsis basicola*. Phytopathology 56:337-342.

Nag Raj, T. R., and Kendrick, B. 1975. A Monograph of *Chalara* and Allied Genera. Wilfrid Laurier University Press, Waterloo, Ontario, Canada.

Rapetti, S. 1982. Gerbera coltura da vaso. Malattie fungine. Clamer Informa 7:43-53.

Scholten, G. 1964. *Nectria radicicola* en *Thielaviopsis basicola* als parasieten van *Cyclamen persicum*. Neth. J. Plant Pathol. Suppl. 2.

Yarwood, C. E. 1946. Isolation of *Thielaviopsis basicola* from soil by means of carrot discs. Mycologia 38:346-348.

Minor Root Diseases

Colletotrichum coccodes (Wallr.) S. J. Hughes, *Cylindrocladiella peruviana* (Batista, Bezerra, & Herrera) Boesewinkel, *Cylindrocladium scoparium* Morg., *Myrothecium roridum* Tode:Fr., and *Sclerotium rolfsii* Sacc. (anamorph of *Athelia rolfsii* (Curzi) Tu & Kimbrough) are soilborne fungi that have been reported to cause root or crown rot on one or more flowering potted plants (Table 9; Fig. 24; Plates 40, 41, and 79). They occur relatively infrequently in the greenhouse but may be brought in via field soil, dirty tools, or plant material.

When field soil is a component of the growing medium, it must be steam treated or fumigated. Fungicides may assist in disease management; however, because of the infrequent occurrence of these minor root rot pathogens in potted plants, application of fungicides as a protectant strategy for their control is rarely warranted.

Diseases Caused by Oomycetes

Oomycetes are an economically important group of fungal plant pathogens, particularly *Pythium*, *Phytophthora*, *Albugo*, and the downy mildews. *Pythium* and *Phytophthora* are distinguished from each other by differences in zoosporogenesis. The downy mildews differ from *Pythium* and *Phytophthora* in sporangial characteristics and because they are obligate parasites. *Albugo* is also an obligate parasite and causes diseases known as white rusts. Oomycete-incited diseases are exacerbated by abundant moisture. The degree of host specialization varies considerably, depending on the species. As a group, they are similar in their sensitivity to certain classes of fungicides.

Pythium is probably the most common plant pathogen associated with root diseases of potted plants. The genus is ubiquitous in soils and has a worldwide distribution. Virulence varies according to the susceptibility and nutritional status of the host, the amounts of moisture and oxygen in the root zone, temperature, and the species of *Pythium*. Table 10 lists *Pythium* species reported on potted plants.

The genus *Phytophthora* contains more than 60 recognized species, including many important plant pathogens. At least 10 species have been reported on potted plants covered in this

TABLE 10. *Pythium* spp. that Cause Diseases of Flowering Potted Plants

Host	Species
Aquilegia sp. (columbine)	*P. debaryanum, P. mamillatum*
Begonia sp.	*P. debaryanum, P. intermedium, P. irregulare, P. splendens, P. ultimum*
Bougainvillea sp.	*P. splendens.*
Calceolaria Herbeohybrida group (pocketbook plant)	*P. ultimum, P. mastophorum*
Capsicum annuum (ornamental pepper)	*P. aphanidermatum, P. irregulare*
Catharanthus roseus (vinca)	*Pythium* sp.
Crossandra infundibuliformis	*Pythium* sp.
Dahlia hybrids	*P. acanthicum, P. debaryanum, P. oedochilum, P. ultimum*
Euphorbia pulcherrima (poinsettia)	*P. aphanidermatum, P. debaryanum, P. perniciosum, P. oligandrum, P. polymastum, P. ultimum*
Eustoma grandiflorum (lisianthus)	*Pythium* sp.
Exacum affine (Persian violet)	*Pythium* sp.
Fuchsia × hybrida	*P. rostratum, P. ultimum*
Gardenia augusta	*P. spinosum, P. splendens*
Gerbera jamesonii (African daisy)	*Pythium* sp.
Hibiscus rosa-sinensis	*Pythium* sp.
Hydrangea macrophylla	*Pythium* sp.
Impatiens sp.	*P. aphanidermatum, P. debaryanum, P. irregulare, P. paroecandrum, P. spinosum, P. ultimum*
Kalanchoe blossfeldiana	*Pythium* sp.
Mimulus × hybridus (monkey flower)	*Pythium* sp.
Pelargonium spp. (geranium)	*P. aphanidermatum, P. debaryanum, P. intermedium, P. mamillatum, P. megalacanthum, P. splendens, P. ultimum, P. vexans*
Pericallis × hybrida (cineraria)	*P. ultimum*
Primula spp. (primrose)	*P. irregulare, P. megalacanthum, P. spinosum, P. ultimum*
Ranunculus sp.	*P. debaryanum, P. irregulare, P. sylvaticum, P. ultimum*
Saintpaulia ionantha (African violet)	*P. spinosum, P. splendens, P. ultimum*
Schlumbergera truncata (Thanksgiving cactus)	*P. aphanidermatum, P. irregulare*
Sinningia speciosa (gloxinia)	*P. debaryanum*
Solanum pseudocapsicum (Jerusalem cherry)	*P. ultimum*

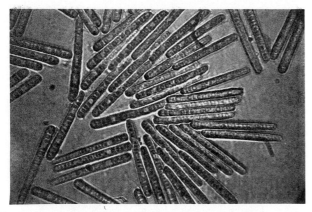

Fig. 24. Two-celled spores of *Cylindrocladium scoparium*.

compendium (Table 11). In general, *Phytophthora* is more pathogenic than *Pythium*, but as with *Pythium*, pathogenicity is dependent on environmental conditions, the host, and the species of *Phytophthora*. The host range varies considerably depending on the species. *P. capsici* Leonian (Fig. 25) has a relatively narrow host range but attacks plants in two different families, Solanaceae and Cucurbitaceae. *P. cinnamomi* Rands

TABLE 11. *Phytophthora* spp. that Cause Diseases of Flowering Potted Plants

Host	Species
Anemone coronaria	*P. cactorum*
Aquilegia sp. (columbine)	*P. drechsleri, P. hibernalis*
Begonia sp.	*P. cactorum, P. cryptogea, P. drechsleri, P. parasitica*
Bougainvillea sp.	*P. nicotianae* var. *parasitica, P. parasitica*
Calceolaria Herbeohybrida group (pocketbook plant)	*P. cinnamomi, P. cryptogea*
Capsicum annuum (ornamental pepper)	*P. capsici, P. n. parasitica*
Catharanthus roseus (vinca)	*P. cactorum, P. colocasiae, P. lateralis, P. nicotianae, P. n. parasitica, P. parasitica, P. palmivora*
Crossandra infundibuliformis	*Phytophthora* sp.
Euphorbia pulcherrima (poinsettia)	*P. drechsleri, P. n. parasitica, P. parasitica*
Exacum affine (Persian violet)	*P. parasitica*
Fuchsia × hybrida	*P. n. parasitica*
Gardenia augusta	*P. parasitica*
Gerbera jamesonii (African daisy)	*P. cryptogea, P. n. parasitica*
Hibiscus rosa-sinensis	*P. cactorum, P. cinnamomi, P. n. nicotianae, P. n. parasitica, P. parasitica*
Impatiens spp.	*Phytophthora* sp.
Kalanchoe blossfeldiana	*P. cactorum, P. n. nicotianae*
Pelargonium spp. (geranium)	*P. parasitica*
Pericallis × hybrida (cineraria)	*P. cambivora, P. cinnamomi, P. cryptogea, P. parasitica*
Primula spp.	*P. cactorum, P. citricola, P. parasitica, P. primulae, P. verrucosa*
Saintpaulia ionantha (African violet)	*P. cactorum, P. n. nicotianae, P. n. parasitica*
Schizanthus sp. (butterfly flower)	*P. cactorum, P. cinnamomi, P. cryptogea*
Schlumbergera truncata (Thanksgiving cactus)	*P. n. parasitica*
Sinningia speciosa (gloxinia)	*P. cactorum, P. cryptogea, P. nicotianae, P. n. nicotianae, P. n. parasitica, P. parasitica*
Solanum pseudocapsicum (Jerusalem cherry)	*P. n. parasitica*

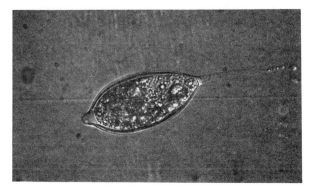

Fig. 25. Papillate sporangium of *Phytophthora capsici.*

attacks more than 900 different hosts. Host ranges for many *Phytophthora* species are not completely known. On flowering potted plants, *P. parasitica* and *P. cryptogea* are the most common species encountered. *P. cactorum, P. cambivora, P. cinnamomi, P. citricola, P. capsici, P. colocasiae, P. drechsleri, P. hibernalis, P. palmivora,* and *P. primulae* have been reported less frequently.

Phytophthora, like *Pythium,* produces oospores and zoospores. It is a facultative parasite; is stimulated by root exudates; and is more pathogenic under conditions of high soil moisture, low oxygen, and high salt levels. Some *Phytophthora* species produce sporangia on aboveground portions of plants and thus can be easily dispersed by splashing water.

Selected References

Dick, M. W. 1990. Keys to *Pythium.* University of Reading, Reading, England.

Erwin, D. C., Bartnicki-Garcia, S., and Tsao, P. H., eds. 1983. *Phytophthora.* Its Biology, Taxonomy, Ecology, and Pathology. American Phytopathological Society, St. Paul, MN.

Middleton, J. D. 1943. The taxonomy, host range and geographic distribution of the genus *Pythium.* Mem. Torrey Bot. Club 20.

Nienhaus, F. 1960. Das Wirtsspektrum von *Phytophthora cactorum* (Leb. et Cohn) Schroet. Phytopathol. Z. 38:33-68.

Stamps, D. J., Waterhouse, G. M., Newhook, F. J., and Hall, G. S. 1990. Revised Tabular Key to the Species of *Phytophthora.* C.A.B. International, Wallingford, England.

Tompkins, C. M. 1975. World literature on *Pythium* and *Rhizoctonia* species and the diseases they cause. Contribution 24. Reed Herbarium, Baltimore, MD.

Van Der Plaats-Niterink, A. J. 1981. Monograph of the Genus *Pythium.* Centraalbureau Voor Schimmelcultures and Institute of the Royal Netherlands Academy of Sciences and Letters, Baarn, Netherlands.

Waterhouse, G. M. 1968. The Genus *Pythium* Pringsheim. Common wealth Mycological Institute, Kew, England.

Waterhouse, G. M. 1970. The Genus *Phytophthora* DeBary. Comonwealth Mycological Institute, Kew, England.

Diseases Caused by *Pythium* spp.

Symptoms

Pythium root and stem rot typically results in yellowing, stunting, or wilt (Plate 80) of the aboveground portions of the plant. Roots may have discrete lesions, and lateral roots and root tips may be decayed. However, often large portions of the root system are decayed (Plate 81). Depending on the host and the stage of disease development, root lesions may appear water soaked, glassy, or brown and shriveled. The decayed cortex is easily stripped off, leaving the vascular strand intact. The fungus may move from the root system several centimeters into the stem, resulting in stem or crown rot (Plate 82) or canker (Plates 83 and 84). In some cases, the bases of unrooted cuttings may be infected. Stem cankers occasionally form in the absence of extensive root rot (Plate 85).

These symptoms are not unique to *Pythium* diseases. Wilting, yellowing, and stunting can be caused by abiotic factors, and root and stem lesions can be caused by a number of other plant-pathogenic fungi. Feeding by fungus gnat larvae may cause symptoms similar to those of *Pythium* infection or may aggravate symptoms on a *Pythium*-infected plant (Plate 86).

Causal Organisms

The genus *Pythium* contains about 125 species, most of which are nonselective regarding the hosts they attack. They are widely distributed in soils. Some species, such as *P. ultimum* Trow, *P. irregulare* Buisman, and *P. aphanidermatum* (Edson) Fitzp., are particularly common as pathogens of

distinct species.

It has been proposed that isolates that apparently attack only begonia be considered a forma specialis, *P. cryptogea* f. sp. *begoniae*.

Host Range and Epidemiology

P. cryptogea has been reported on the potted plants *Begonia* spp., *Calceolaria* Herbeohybrida group (pocketbook plant), *Pericallis* × *hybrida* R. Nordenstam (cineraria) (Plate 87), *Gerbera jamesonii* H. Bolus ex Adlam, *Schizanthus* × *wisetonensis* hort. (butterfly flower), and *S. speciosa*. It has also been reported on other hosts, including *L. esculentum*, *Solanum tuberosum* L. (potato), *Cichorium intybus* L. (chicory), *Spinacia oleracea* L. (spinach), *Phaseolus vulgaris* L. (bean), *Allium cepa* L. (onion), cucurbits, and *Solanum melongena* L. var. *esculentum* Nees. (eggplant).

The optimum temperature for growth of *P. cryptogea* is 22–25°C. As is generally true for other *Phytophthora* species, zoospore formation is stimulated by saturated soil. Sporangia form at temperatures of 5°C to nearly 30°C; the optimum is approximately 20°C. Root exudates stimulate zoospore development. Typically the roots are invaded, but the crown of the plant may become colonized as well.

Management

Management practices outlined for *Pythium* are also applicable to *P. cryptogea*.

Selected References

Kaewruang, W., Sivasithamparam, K., and Hardy, G. E. 1988. *Phytophthora cryptogea*, an additional pathogen of African daisy in western Australia. Australas. Plant Pathol. 17:67-68.

MacDonald, J. D., and Duniway, J. M. 1978. Temperature and water stress effects on sporangium viability and zoospore discharge in *Phytophthora cryptogea* and *P. megasperma*. Phytopathology 68:1449-1455.

Middleton, J. T., Tucker, C. M., and Tompkins, C. M. 1944. A disease of gloxinia caused by *Phytophthora cryptogea*. J. Agric. Res. 68:405-413.

Orlikowski, L. 1979. Effect of temperature, pH and peat moisture on sporulation of *Phytophthora cryptogea* from diseased African daisy. Bull. Acad. Pol. Sci. 27:761-767.

Orlikowski, L. 1981. Studies on the biological control of *Phytophthora cryptogea* Pethybr. et Laff., the mycoflora associated with African daisy production in Polish greenhouses and effects of its main components on the development of the pathogen. Prot. Ecol. 2:285-296.

Rattink, H. 1981. Characteristics and pathogenicity of six *Phytophthora* isolates from pot plants. Neth. J. Plant Pathol. 87:83-90.

Stamps, D. J. 1978. *Phytophthora cryptogea*. Descriptions of Pathogenic Fungi and Bacteria, no. 592. Commonwealth Mycological Institute, Kew, England.

Von Krober, H. 1981. Vergleichende untersuchungen vom grundtyp *Phytophthora cryptogea* und *P. drechsledri* abweichender isolate. Phytopathol. Z. 102:219-231.

Yoshimura, M. A., Uchida, J. Y., and Aragaki, M. 1985. Etiology and control of poinsettia blight caused by *Phytophthora nicotianae* var. *parasitica* and *P. drechsleri*. Plant Dis. 69:511-513.

Diseases Caused by *Phytophthora parasitica*

Phytophthora parasitica is the most commonly encountered *Phytophthora* species in potted plants. Typically roots, crowns, or stems are invaded, but occasionally foliar blights occur. *P. parasitica* is considered to be the most important soilborne pathogen of *Saintpaulia ionantha* Wendl. (African violet), *Sinningia speciosa* (Lodd.) Hiern (gloxinia), and *Catharanthus roseus* (L.) G. Don (vinca). As with other *Phytophthora* spp., the adoption of soilless potting media has reduced the importance of this pathogen.

Symptoms

On African violet and gloxinia, symptoms caused by *P. parasitica* root and crown rot (Plates 88 and 89) are similar to those caused by *Pythium* and may also be confused with symptoms of impatiens necrotic spot tospovirus. The first symptom is usually a wilting of one or more leaves. Roots become water soaked and brown to black, and the cortex is easily removed. A brown to black lesion may also be apparent at the crown of the plant, from which colonization of petioles and leaves may occur. Leaves may also become colonized from contact with the growing medium, resulting in a water-soaked lesion that becomes brown and flaccid. The corm of gloxinia may also become invaded and rot.

Holiday cacti infected with *P. parasitica* show darkening of cladophylls at the soil line followed by wilting of the affected stem. The interior of the stem may have an orange discoloration in the area of the decay (Plate 90).

All plant parts of poinsettia (*Euphorbia pulcherrima* Willd. ex Klotzsch) are susceptible. On stems, wet, gray lesions may develop, or dark, elongated streaks may originate at the crown (Plate 91). On woody stems, crown discoloration may not be apparent externally (Plate 92). The pith becomes brown, the discoloration extending beyond the canker. Individual stems may become colonized and die. Leaves develop small, angular, tan spots that may coalesce into large areas of necrosis (Plate 93). Leaves and petioles of infected terminals bend downward sharply. Bracts are of poor quality on infected plants (Plate 94).

Vinca may develop damping-off, leaf blight, or stem canker. Damping-off is particularly severe in plug trays (Plate 95) where water splash can disseminate zoospores. An increase in branching of the infected roots may be noted. Aerial blight may occur in the absence of root rot. Leaves first become gray and flaccid, and then necrosis occurs (Plate 96). A lesion can begin at the growing tip and move down the stem, or it may originate as a canker.

A foliage blight of *Bougainvillea* caused by *P. parasitica* was described in Florida in 1970. Initial lesions are small, irregular, and gray green and often occur on the tips and margins of young leaves. Lesions enlarge rapidly, and the foliage becomes flaccid, blackened, and curled. Mature leaves are reported to be resistant to infection. *Phytophthora* may move from the leaves through the petiole and into the stem where extensive blighting may occur. Under conditions of high humidity, mycelium may develop on the plant surface.

Gardenia (*Gardenia augusta* (L.) Merrill) is susceptible to Phytophthora root rot (Plate 97). Aboveground symptoms are not apparent until a significant portion of the root system decays (Plate 98). Leaves become chlorotic and abscise. When crown rot occurs, individual branches may die. One study indicated considerable variation in pathogenicity of *P. parasitica* isolates to gardenia.

Causal Organism

P. nicotianae Breda de Haan was first described from *Nicotiana tabacum* L. (tobacco) in 1896. In 1913, *P. parasitica* Datsur was described. Waterhouse combined the two species into *P. nicotianae* var. *parasitica* and *P. nicotianae* var. *nicotianae*, distinguished by morphological and pathological characteristics. Recent evidence, including mitochondrial DNA analysis, isozyme analysis, and morphological comparisons do not support the concept of two separate species or two varieties; however, isolates identified as *P. n. nicotianae* may include strains that are restricted to tobacco. On the other hand, *P. n. nicotianae*, as identified by morphological features, has been recorded on *Peperomia, Saintpaulia, Hibiscus, Kalanchoe,* and *Sinningia,* all of which are outside the Solanaceae. There is considerable physiological and pathological variation in isolates of *P. parasitica.* For example, highly virulent isolates from gloxinia are not necessarily highly virulent on African violet. In this compendium, *P. n. nicotianae* and *P. n. parasitica* are referred to as *P. parasitica,* except in Table 11, where the original designation used by the author is retained.

P. parasitica has prominently papillate sporangia and amphigynous antheridia, placing it in Waterhouse's group II. *P. parasitica* is self-incompatible, although some isolates will form oospores when cultured from plants, presumably because they are mixtures of both mating types. Oospores are typically less than 20 μm in diameter, although they may be larger in some isolates. Sporangia are mostly spherical to ovoid with a length-breadth ratio of less than 1.6. (Fig. 31). *P. parasitica* has a fairly high optimum temperature range (25–32°C) for growth. The ability to grow at 35°C and its distinctive patchy growth pattern on corn meal agar distinguish it from other *Phytophthora* species (Fig. 32).

Host Range and Epidemiology

P. parasitica attacks plants in more than 58 different families. Flowering potted plants reported as hosts include *Begonia, Bougainvillea, Capsicum annuum* L. var. *annuum* (ornamental pepper), *C. roseus, E. pulcherrima, Exacum affine* Balf. f., *Fuchsia* × *hybrida* Hort. ex Vilm. (Plate 99), *G. augusta, Gerbera jamesonii* H. Bolus ex Adlam (African daisy) (Plate 100), *Hibiscus rosa-sinensis* L., *Kalanchoe blossfeldiana* Poelln., *Lantana camara* L., *S. ionantha, Schlumbergera truncata* (Haw.) Moran (Thanksgiving cactus), *Sinningia speciosa* (Lodd.) Hiern (gloxinia), and *Solanum pseudocapsicum* L. (Jerusalem cherry). Other flowering potted plants may also be hosts. Cultivars of African violet, gloxinia, vinca and *Bougainvillea* vary in their susceptibility to *P. parasitica.*

Rapid disease development occurs at temperatures above 28°C, but colonization in the absence of symptoms may occur at lower temperatures. Under saturated conditions, zoospores are released and disseminated by splashing and running water. The fungus is soilborne and largely dependent on the movement of contaminated soil, water, and plant material for dispersal. Circumstantial evidence indicates that the fungus may be seedborne in *C. roseus.*

Management

Management practices (including effective fungicides) outlined for *Pythium* are also applicable to *P. parasitica.* A metalaxyl-insensitive isolate of *P. parasitica* has been reported from California.

Selected References

Alfieri, S. A., Jr. 1970. Bougainvillea blight, a new disease caused by *Phytophthora parasitica.* Phytopathology 60:1806-1808.

Alfieri, S. A., Jr. 1970. Crown rot of gloxinia. Fla. Dep. Agric. Consumer Serv. Div. Plant Indus. Plant Pathol. Circ. 96.

Busch, L. V., and Smith, E. A. 1978. Control of root and crown rot of African violet and of gloxinia by *Phytophthora nicotianae* var. *nicotianae.* Can. Plant Dis. Surv. 58:73-74.

Daughtrey, M., and Macksel, M. 1986. Efficacy of fosetyl-Al and metalaxyl against *Phytophthora parasitica* on poinsettia. (Abstr.) Phytopathology 76:651.

Engelhard, A. W., and Ploetz, R. C. 1979. Phytophthora crown and stem rot, an important new disease of poinsettia (*Euphorbia pulcherrima*). Proc. Fla. State Hortic. Soc. 92:348-350.

English, J. T., and Mitchell, D. J. 1984. Patterns of root growth and inoculum production in the *Catharanthus roseus-Phytophthora parasitica* pathosystem. (Abstr.) Phytopathology 74:813.

Ferrin, D. M., and Rohde, R. G. 1992. In vivo expression of resistance to metalaxyl by a nursery isolate of *Phytophthora parasitica* from *Catharanthus roseus.* Plant Dis. 76:82-84.

Holcomb, G. E. 1993. First report of flower blight of lantana caused by *Phytophthora parasitica.* Plant Dis. 77:1168.

Keim, R. 1977. Foliage blight of periwinkle in southern California. Plant Dis. Rep. 61:182-184.

Krober, H., and Plate, H.-P. 1973. *Phytophthora*-Faule an *Saintpaulien* [Erregerf: *Phytophthora nicotianae* var. *parasitica* (Dast.) Waterh.]. Phytopathol. Z. 76:348-355.

Oudemans, P., and Coffey, M. D. 1991. A revised systematics of twelve papillate *Phytophthora* species based on isozyme analysis. Mycol. Res. 95:1025-1046.

Ploetz, R. C., and Engelhard, A. W. 1980. Crown and root rot of gloxinia and other gesneriads caused by *Phytophthora parasitica.* Plant Dis. 64:487-490.

Rattink, H. 1981. Characteristics and pathogenicity of six *Phytophthora* isolates from pot plants. Neth. J. Plant Pathol. 87:83-90.

Rattink, H. 1981. *Phytophthora nicotianae* var. *nicotianae* on *Sinningia*: Cultivar reaction and control. Med. Fac. Landbouwwet. Rijksuniv. Gent 46:881-887.

Schubert, T. S., and Leahy, R. M. 1989. Phytophthora blight of *Catharanthus roseus.* Fla. Dep. Agric. Consumer Serv. Div. Plant Indus. Plant Pathol. Circ. 321.

Strider, D. L. 1978. Reaction of African violet cultivars to *Phytophthora nicotianae* var. *parasitica.* Plant Dis. Rep. 62:112-114.

Tsao, P. H., and Sisemore, D. J. 1978. Morphological variability in *Phytophthora parasitica* (*P. nicotianae*) isolates from citrus, tomato, and tobacco. (Abstr.) Phytopathol. News 12:213.

Waterhouse, G. M., and Waterston, J. M. 1964. *Phytophthora nicotianae* var. *nicotianae.* Descriptions of Pathogenic Fungi and Bacteria, no. 34. Commonwealth Mycological Institute, Kew, England.

Waterhouse, G. M., and Waterston, J. M. 1964. *Phytophthora nicotianae* var. *parasitica.* Descriptions of Pathogenic Fungi and Bacteria, no. 35. Commonwealth Mycological Institute, Kew,

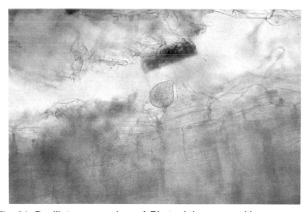

Fig. 31. Papillate sporangium of *Phytophthora parasitica.*

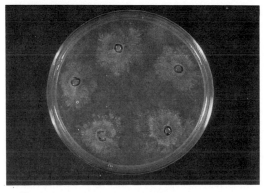

Fig. 32. *Phytophthora parasitica* cultured from a cankered poinsettia stem.

England.

Wick, R. L. 1993. *Phytophthora parasitica* root rot of *Gardenia*. Mass. Flower Grow. Assoc. Res. Rep. 4.

Yoshimura, M. A., Uchida, J. Y., and Aragaki, M. 1985. Etiology and control of poinsettia blight caused by *Phytophthora nicotianae* var. *parasitica* and *P. drechsleri*. Plant Dis. 69:511-513.

White Rusts

White rusts are obligate pathogens that are closely related to the downy mildews. They are not true rusts but are so named because they form white, rustlike pustules on various plant parts. White rusts are of minor economic importance on potted plants. *Gerbera jamesonii* H. Bolus ex Adlam (African daisy) and *Pericallis × hybrida* R. Nordenstam (cineraria) are susceptible to a white rust fungus.

Symptoms

White rust on African daisy first appears as light spots on the lower surfaces of leaves. Raised pustules then develop beneath the epidermis, forming blisters that burst with the pressure that results from sporangiospore production. A plant infected with white rust shows white, crusty masses (pustules with exposed sporangiospores) on the undersurfaces of leaves (Fig. 33).

Causal Organism

White rust of African daisy and cineraria is caused by *Albugo tragopogonis* (Pers.) S. F. Gray, a member of the Albuginaceae. Unlike other oomycetes, *Albugo* produces chains of sporangiospores (18–25 × 11–15 μm) that are cylindrical to cuboid and hyaline to pale yellow. Oospores (40–65 μm in diameter) are dark brown with spiny, reticulate, or tuberculate walls. Sporangiospores are spread by wind currents to nearby plants and germinate by forming zoospores, which initiate new infections.

Host Range and Epidemiology

A. tragopogonis causes white rust on plants within the aster family. Flowering potted plant hosts include cineraria and African daisy (gerbera). Although this disease (described as white blister) is common on various species of *Pericallis* and affects greenhouse crops of dusty miller, it is not common on *P. × hybrida* and has not been reported on this flowering potted plant in the United States. White rust on African daisy is common in Australia and New Zealand but has not been reported in the United States.

Infection of host plants usually occurs via sporangiospores, the disseminative spore form of *A. tragopogonis*. The sporangiospores produce motile zoospores that encyst, germinate, and penetrate the host epidermis via a germ tube. An intercellular mycelium then forms within host tissue, absorbing nutrients through intracellular, knoblike haustoria. After a short period, clusters of short, club-shaped sporangiophores are produced at the tips of hyphal strands. The sporangiospores are formed in chains on each sporangiophore. At maturity, these spores burst through the host epidermis to form soruslike fruiting structures that resemble the pustules produced by the true basidiomycete rusts.

Oospores form in the outer cells of the host plant and allow *A. tragopogonis* to resist adverse environmental factors such as desiccation and low temperatures. *A. tragopogonis* is not a soilborne organism; it survives in the greenhouse as sporangiospores or oospores on dormant plants and debris. During moist, warm weather, oospores germinate to form vesicles that contain several zoospores capable of initiating new infections.

Management

To effectively reduce disease incidence, diseased plants should be discarded and the introduction of diseased plants into the greenhouse should be avoided.

Selected References

Doepel, R. F. 1965. White rust of African daisies. J. Agric. West. Aust. Ser. 4 6(7):439. (Rev. Appl. Mycol. 45:80, 1966)

Plate, H. P., and Krober, H. 1977. Weisser Rost an Gerbera auf Teneriffa (Erreger: *Albugo tragopogonis* (DC.) S. F. Gray). (White rust on Gerbera in Tenerife (causal agent: *Albugo tragopogonis* (DC.) S. F. Gray) Nachr. Deut. Pflanzenschutzd. 29(11):169-170. (Rev. Plant Pathol. 57:204, 1978)

Downy Mildews

The downy mildews are represented by six genera, but only *Peronospora*, *Plasmopara*, and *Bremia* have been reported on the plants covered in this compendium (Table 12). Downy mildews are not a common problem on potted plants.

The most important characteristic that distinguishes the downy mildews from other oomycetes is that they are obligate parasites. The downy mildews also have determinate sporangiophores, while *Phytophthora* and *Pythium* have indeterminate sporangiophores. The genera of downy mildews are distinguished largely by the morphology of their sporangia. In addition, *Peronospora* does not produce zoospores; the sporangia germinate directly. *Bremia* may or may not produce zoospores, and *Plasmopara* does produce zoospores.

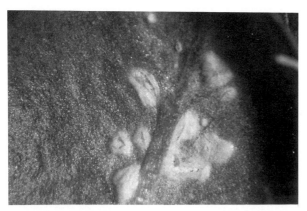

Fig. 33. Spore pustules of white rust (*Albugo tragopogonis*) on the underside of a leaf.

TABLE 12. Downy Mildew Fungi that Affect Flowering Potted Plants

Host	Fungus
Anemone coronaria	*Plasmopara pygmaea*
Capsicum annuum (ornamental pepper)	*Peronospora tabacina*
Eustoma grandiflorum (lisianthus)	*Peronospora chlorae*
Gerbera jamesonii (African daisy)	*Plasmopara* sp.
Impatiens spp.	*Plasmopara obducens*
Mimulus sp. (monkey-flower)	*Peronospora jacksonii*
Pericallis × hybrida (cineraria)	*Plasmopara halstedii, Bremia lactucae, Peronospora gangliformis*
Primula spp. (primrose)	*Peronospora oerteliana*
Ranunculus sp.	*Peronospora ficariae, P. hiemalis, P. pennsylvanica, P. ranunculi*

A downy mildew of cineraria (*Pericallis × hybrida* R. Nordenstam) caused by *Plasmopara halstedii* (Farl.) Berl. & De Toni in Sacc. was reported in 1933 when it caused significant losses in Long Island, New York. The disease has also been reported from the Pacific Northwest. Symptoms of this disease are first evident as white spots 0.5–3 cm in diameter on the undersides of the leaves. The spots become dark brown, and the entire leaf dies.

High humidity, prolonged leaf wetness, and cool weather are conducive to development of downy mildew diseases. Optimum temperatures for downy mildew fungi vary from 15 to 23°C for *Peronospora tabacina* D. B. Adam and from 6 to 20°C for *Bremia lactucae* Regel. Once infection occurs, sporangiophores emerge through the stomata on the underside of the leaf (Fig. 34). Sporangia soon follow and are easily dislodged and disseminated by wind and splashing water. An infected leaf often exhibits chlorotic patches on the upper surface opposite fuzzy, grayish areas of sporulation on the leaf underside (Plates 101–103).

Downy mildew fungi may reside in infected plant debris in the soil or on weed hosts. When outbreaks occur, diseased plants and weeds should be removed. Conditions that allow for periods of prolonged leaf wetness should be avoided. Fungicides effective against oomycetes, including mancozeb, fosetyl Al, and metalaxyl, have shown effectiveness against downy mildews. The possibility that a downy mildew may develop resistance to metalaxyl is a concern; tank mixes with mancozeb have been used in vegetable crop systems to prevent this.

Selected References

Gill, D. L. 1933. *Plasmopara halstedii* on cineraria. Mycologia 25:446-447.

Spencer, D. M. The Downy Mildews. 1981. Academic Press, London.

Powdery Mildew Diseases

A powdery mildew fungus is an obligate parasite and thus requires a living plant host to complete its life cycle. It does not usually kill its host, but powdery mildew colonies quickly render a flowering potted plant unmarketable. The powdery mildews infect a wide range of plants and are well represented on ornamentals. Powdery mildews are particularly important on *Saintpaulia ionantha* Wendl. (African violet), *Begonia*, *Dahlia*, *Gerbera jamesonii* H. Bolus ex Adlam (African daisy), *Hydrangea*, *Euphorbia pulcherrima* Willd. ex Klotzsch (poinsettia), and *Kalanchoe blossfeldiana* Poelln.

Potted plants covered in this compendium are host to the powdery mildew genera *Erysiphe*, *Leveillula*, *Microsphaera*, *Sphaerotheca*, and *Uncinula*, as well as the form genus *Oidium*. Each fungus forms a network of hyphae over the plant surface from which it penetrates epidermal cells, deriving nutrients via haustoria (Fig. 35).

Symptoms

Powdery mildew genera cannot be distinguished on the basis of symptoms. They all produce a superficial vegetative growth of hyphae on leaves and often on stems and flowers. Conidia (Fig. 36) are borne singly or in chains (Fig. 37) on upright conidiophores arising from the mycelium. These spores contribute to the white, powdery appearance of the fungus on the plant surface (Plates 104 and 105). The disease is usually easy to identify from its conspicuous growth on the upper leaf surface; in some cases, however, the fungus may cause necrosis to an area on the upper leaf surface with little if any discernible mycelium (Plate 106). In other cases, the upper leaf surface may have patches of chlorosis or necrosis

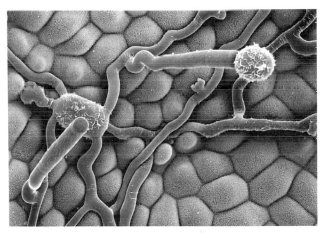

Fig. 35. Powdery mildew colony on the surface of a poinsettia leaf. (Courtesy G. Celio)

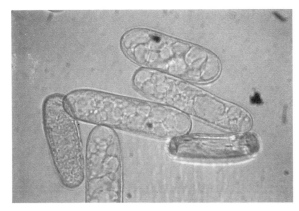

Fig. 34. Sporangiophores and sporangia of a *Peronospora* sp. (downy mildew). (Courtesy T. Bowyer)

Fig. 36. Large, elongated conidia of *Microsphaera begoniae*.

Gronberg, H. 1987. Mildew on Ranunculus can be controlled. Gartner Tidende 103:713.

Hausbeck, M., Kalishek, J., Daughtrey, M., and Barnes, L. 1994. Keep the colonies at bay, part 1. Greenhouse Grower 12(9):45, 47-48.

Hausbeck, M., Kalishek, J., Daughtrey, M., and Barnes, L. 1994. Keep the colonies at bay, part 2. Greenhouse Grower 12(10):88, 91-92.

Hou, H. H., and Lee, C. S. 1979. Powdery mildew (*Sphaerotheca fuliginea*) of *Dahlia pinnata* in Taiwan. Plant Prot. Bull. 21:441-443.

Ing, B. 1990. An introduction to British powdery mildews—1–4. Mycologist 4:46-48, 88-90, 125-128, 172-177.

Ing, B. 1991. An introduction to British powdery mildews—5. Mycologist 5:24-27.

Kapoor, J. N. 1967. *Erysiphe cichoracearum*. Descriptions of Pathogenic Fungi and Bacteria, no. 152. Commonwealth Mycological Institute, Kew, England.

Kapoor, J. N. 1967. *Sphaerotheca fuliginea*. Descriptions of Pathogenic Fungi and Bacteria, no. 159. Commonwealth Mycological Institute, Kew, England.

Kontaxis, D. G. 1985. Managing powdery mildew on begonia. Calif. Agric. 39:16.

Peterson, J. L., and Davis, S. H., Jr. 1970. Suppression of *Erysiphe polygoni* and *Botrytis cinerea* on hydrangea with benomyl. Plant Dis. Rep. 54:606-607.

Powell, C. C. 1993. Controlling powdery mildew on poinsettias by controlling leaf wetness and using fungicides. Pages 3-5 in: Ohio Florists Assoc. Bull. 759.

Quinn, J. A., and Powell, C. C., Jr. 1981. Identification and host range of powdery mildew of begonia. Plant Dis. 65:68-70.

Quinn, J. A., and Powell, C. C., Jr. 1982. Effects of temperature, light, and relative humidity on powdery mildew of begonia. Phytopathology 72:480-484.

Quinn, J. A., and Powell, C. C., Jr. 1982. Effectiveness of new systemic fungicides for control of powdery mildew of begonia. Plant Dis. 66:718-720.

Spencer, D. M. 1978. The Powdery Mildews. Academic Press, New York.

Strider, D. L. 1974. Resistance of Rieger elatior begonias to powdery mildew and efficacy of fungicides for control of disease. Plant Dis. Rep. 58:875-878.

Strider, D. L. 1980. Resistance of African violet to powdery mildew and efficacy of fungicides for control of the disease. Plant Dis. 64:188-190.

Strider, D. L. 1983. Efficacy of fungicides for the eradication of powdery mildew of Kalanchoe. Fungic. Nematicide Tests 38:177.

Viennot-Bourgin, G. 1983. Les oidiums du begonia. Cryptogam. Mycol. 4:179-187.

Zaracovitis, C. 1965. Attempts to identify powdery mildew fungi by conidial characteristics. Trans. Br. Mycol. Soc. 48:553-558.

Rust Diseases

Rust diseases, best known for devastating cereal crops, affect a wide range of host plants. The economic importance of these fungal diseases on ornamental potted plants is not usually great; however, geranium rust can cause significant losses. The rusts occurring on potted plants are classified as basidiomycetes in the order Uredinales and are in the family Pucciniaceae or Melampsoraceae.

The rusts are obligate parasites and are usually host specific or have limited host ranges. If several hosts are infected by a given rust, they are usually from the same plant family. Rusts are identified mainly by the type of spores produced, the spore characteristics, and the host plant infected.

The life cycle of rusts can be complex; in some cases, as many as five different spore types are produced. Autoecious rusts need only one host species to complete their life cycle and produce all spore stages on that host. Heteroecious rusts

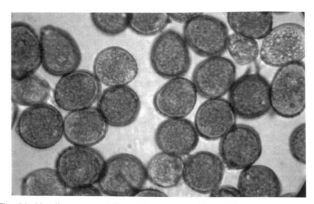

Fig. 38. Urediospores of *Puccinia pelargonii-zonalis*.

require two different host genera to complete their life cycle. For example, they typically produce two spore types on an herbaceous host and two additional spore types on a very different host plant, such as a woody or grass species. Regardless of the economic importance of the hosts, the one on which teliospores are produced is designated the primary host and the other is considered the alternate host.

The five different spore types produced by rusts are designated 0 (pycnial), I (aecial), II (uredial), III (telial), or IV (basidial). Each of the five spore types has a specific function, but only three (aecial, uredial, and basidial) are involved in the infection of a host plant.

Urediospores, when present in the rust life cycle, can reinfect the same host plant and play a dominant role in both dissemination and disease severity (Fig. 38). These spores are generally produced throughout the growing season of the host plant. They are often referred to as the repeating spore stage and are generally responsible for the secondary spread of the rust. The uredial stage is generally responsible for extensive injury to greenhouse crops. When urediospores are missing in a rust cycle, aeciospores often assume their role and are responsible for disease increase. Teliospores are important for overwintering of some rusts and also give rise to basidiospores. In heteroecious rusts, basidiospores infect the alternate host.

Rust fungi survive as spores, as systemic mycelium on or in dormant plants, or in plant debris in the greenhouse. Spores are spread by air currents and can be introduced into a greenhouse on diseased plants. Among the flowering potted plants, rusts are found most commonly on geraniums (*Pelargonium* spp.), fuchsia (*Fuchsia* × *hybrida* Hort. ex Vilm.), and anemone (*Anemone coronaria* L.). Table 14 lists rusts reported on potted plants.

Selected References

Arthur, J. C. 1962. Manual of the Rusts in United States and Canada. Hafner, New York.

Cummins, G. B., and Hiratsuka, Y. 1983. Illustrated Genera of Rust Fungi. American Phytopathological Society, St. Paul, MN.

Geranium Rust

Geranium rust was first described in 1926 and observed in the continental United States in 1967. Although it is a potentially devastating disease, state and federal quarantines, sanitation measures, and grower cooperation contribute to its control in greenhouses.

Symptoms

Symptoms first appear as white or yellowish spots on both leaf surfaces. On the lower leaf surface, spots form directly

1. Alternaria leaf spot on a regal geranium *(Pelargonium x domesticum).*

2. Alternaria leaf spot on vinca (*Catharanthus roseus*). (Courtesy D. O. Gilrein)

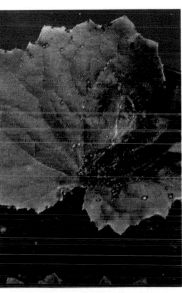

3. Veinal necrosis and leaf blotch caused by *Alternaria euphorbiicola* on poinsettia. (Courtesy A. W. Engelhard)

4. Stem lesion caused by *Alternaria euphorbiicola* on poinsettia. A greasy-appearing dark streak is evident below the lesion. (Courtesy A. W. Engelhard)

5. Purple-bordered leaf lesions caused by an *Alternaria* sp. on cineraria.

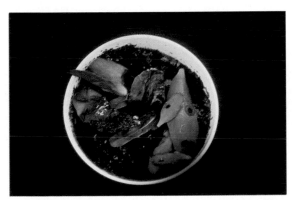

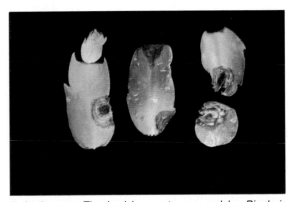

6. Shattered cladophylls of Easter cactus caused by *Bipolaris cactivora*. (Courtesy A. R. Chase)

7. Lesions on Thanksgiving cactus caused by *Bipolaris cactivora*.

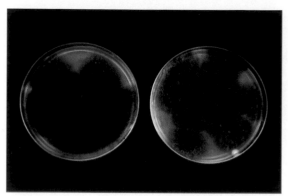

8. Dark mycelium of *Bipolaris cactivora* grown on potato-dextrose agar.

9. Lesions on hydrangea leaves caused by *Botrytis cinerea* during propagation. (Courtesy S. H. Davis)

10. Botrytis crown rot of calceolaria. (Courtesy A. R. Chase)

11. Botrytis canker of lisianthus (*Eustoma grandiflorum*).

12. Botrytis stem canker on vinca (*Catharanthus roseus*). (Courtesy M. A. Hansen)

13. Sporulation of *Botrytis cinerea* on a stem lesion on *Impatiens wallerana*. (Courtesy J. A. Matteoni)

14. Sporulation of *Botrytis cinerea* on hydrangea stems. (Courtesy J. A. Matteoni)

15. Stem rot of kalanchoe, showing sporulation of *Botrytis cinerea*. (Courtesy D. Karasevicz)

16. Tan stem lesion on poinsettia caused by *Botrytis cinerea*.

17. Lesions caused by *Botrytis cinerea* developing at the edge of poinsettia bracts. (Courtesy M. F. Heimann)

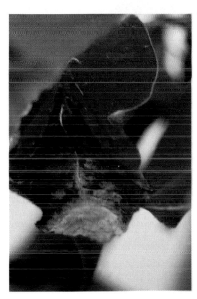

18. *Botrytis cinerea* lesion at the tip of a poinsettia bract.

19. *Botrytis cinerea* sporulation on poinsettia flower parts. (Courtesy M. K. Hausbeck)

20. *Botrytis cinerea* sporulation on tan canker at the base of a Persian violet (exacum) stem. (Courtesy J. A. Matteoni)

21. Dieback of florist's geranium caused by *Botrytis cinerea*. (Courtesy M. K. Hausbeck)

22. Leaf lesions on florist's geranium resulting from *Botrytis cinerea*-infected flowers falling on the foliage.

23. *Botrytis cinerea* sporulation. (Courtesy J. A. Matteoni)

24. Angular leaf spot on hibiscus caused by *Cercospora* sp. (Courtesy A. R. Chase)

25. Leaf spots on hydrangea caused by *Cercospora hydrangeae*. (Courtesy R. K. Jones)

26. Wet rot of poinsettia and sporulation of *Choanephora cucurbitarum*. (Courtesy R. K. Jones)

27. Anemone leaf curl, caused by *Colletotrichum* sp.

28. Anthracnose on cyclamen caused by *Glomerella cingulata*.

29. Crown lesion on cyclamen caused by *Glomerella cingulata*. Note orange sporulation of the anamorph, *Colletotrichum gloeosporioides*, at the base of the pedicel.

30. Petal lesions on cyclamen caused by *Glomerella cingulata*.

31. Constriction and necrosis of young cyclamen flowers and leaves typical of anthracnose caused by *Cryptocline cyclaminis*. (Courtesy J. A. Matteoni)

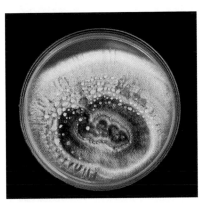

32. *Cryptocline cyclaminis* growing on potato dextrose agar.

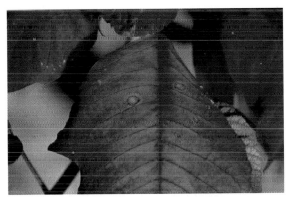

33. Unexpanded lesions on a poinsettia bract caused by *Corynespora cassiicola*. (Courtesy G. W. Simone)

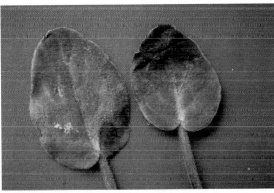

34. Brown leaf lesions on African violet caused by *Corynespora cassiicola*. (Courtesy A. R. Chase)

35. Lesions on dahlia leaves caused by a white smut (*Entyloma calendulae* f. *dahliae*).

36. Leaf lesions on gardenia caused by *Myrothecium roridum*. (Courtesy A. R. Chase)

37. Sporodochia of *Myrothecium roridum* at the base on a Persian violet (exacum) stem. (Courtesy J. A. Matteoni)

38. Leaf lesions on gardenia caused by *Mycosphaerella gardeniae*. (Courtesy G. W. Simone)

39. Brown leaf spot on New Guinea impatiens caused by *Myrothecium roridum*.

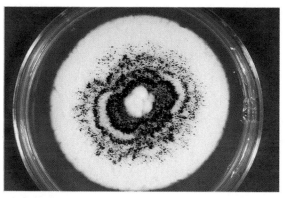

40. Collapse of a young gloxinia plant resulting from crown rot caused by *Myrothecium roridum*. (Courtesy J. A. Matteoni)

41. A 2-wk-old culture of *Myrothecium roridum* on potato-dextrose agar.

42. Cyclamen with stunt disease caused by *Ramularia cyclaminicola* (left) and a healthy plant (right).

43. Ramularia leaf spot of primula. (Copyright J. Fletcher)

46. Rhizopus blight of gloxinia flowers. (Courtesy A. R. Chase)

44. Rhizopus blight of African daisy (gerbera) foliage. (Courtesy A. R. Chase)

45. Sporulation of *Rhizopus stolonifer* at the base of a poinsettia. (Courtesy J. Uchida)

47. Rhizopus stem rot of poinsettia.

48. Rhizopus stem rot of vinca (*Catharanthus roseus*).

49. Poinsettia scab lesions caused by *Sphaceloma poinsettiae*. (Courtesy A. W. Engelhard)

51. Pink sporodochia of *Fusarium solani* and red perithecia of *Nectria haematococca* at the base of a Persian violet (exacum) stem. (Courtesy J. A. Matteoni)

52. Early flowering of Persian violet caused by Nectria canker. (Courtesy J. A. Matteoni)

50. Severe symptoms of scab on a poinsettia stem. (Courtesy A. W. Engelhard)

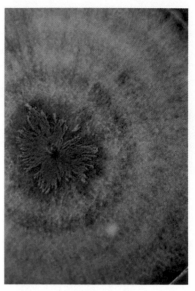

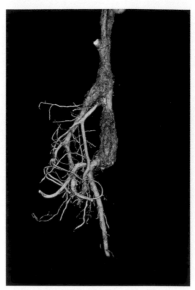

53. Sporulation of *Nectria haematococca* on African daisy (gerbera). (Courtesy J. A. Matteoni)

54. Culture of *Fusarium solani*. (Courtesy J. A. Matteoni)

55. Swelling at the base of a gardenia stem infected by *Phomopsis gardeniae*. (Courtesy G. W. Simone)

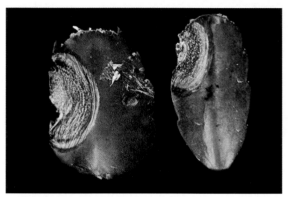

56. Lesions at the bases of African violet petioles caused by *Cylindrocarpon destructans*.

57. Lesions on cladophylls of holiday cactus caused by *Fusarium oxysporum*. (Courtesy A. R. Chase)

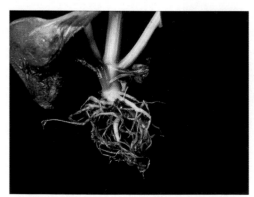

58. Fusarium root rot of *Exacum affine* (Persian violet). (Courtesy W. Jarvis, Agriculture Canada)

59. Basal canker on poinsettia with sporulation of *Fusarium solani* and *Nectria haematococca*. (Courtesy A. W. Engelhard)

60. Web blight of impatiens caused by *Rhizoctonia solani*.

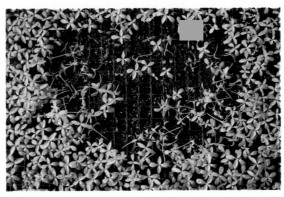

61. Damping-off of vinca (*Catharanthus roseus*) caused by *Rhizoctonia solani*.

62. Rhizoctonia blight of impatiens.

63. Rhizoctonia blight of begonia foliage during propagation. (Courtesy J. A. Matteoni)

64. *Rhizoctonia solani* invading a poinsettia leaf that is in contact with the growing medium.

65. Lesions caused by *Rhizoctonia solani* on leaves of New Guinea impatiens in contact with the growing medium.

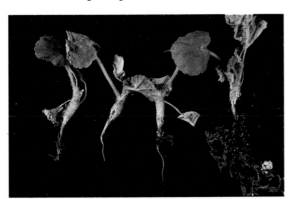

66. Geranium seedlings with Rhizoctonia root rot.

67. Rot at the bases of African violet petioles caused by *Rhizoctonia solani*.

68. Canker at the base of a poinsettia stem caused by *Rhizoctonia solani.*

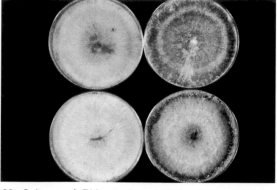

69. Cultures of *Rhizoctonia solani* showing variation in color and sclerotium production.

70. White mycelial tufts of *Sclerotinia sclerotiorum* on a cankered poinsettia stem. (Courtesy A. W. Engelhard)

71. Collapse of poinsettias caused by Thielaviopsis root rot.

72. Wilting of fuchsia caused by Thielaviopsis root rot.

73. Chlorosis of vinca (*Catharanthus roseus*) caused by Thielaviopsis root rot. (Courtesy L. Barnes)

74. Roots of fuchsia blackened by *Thielaviopsis basicola.*

75. Healthy poinsettia roots (right) and roots infected with *Thielaviopsis basicola.*

76. Dark chlamydospores and cracking at the base of a poinsettia stem infected with *Thielaviopsis basicola.*

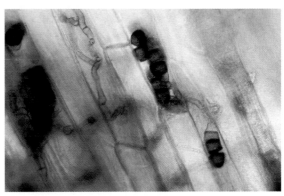

77. Chlamydospores of *Thielaviopsis basicola* within a poinsettia root. (Courtesy M. F. Heimann)

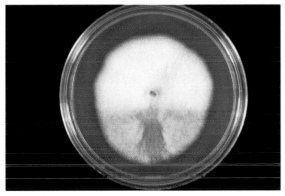

78. *Thielaviopsis basicola* on potato-dextrose agar showing sectoring from the normal, darkly pigmented mycelium.

79. Mycelium of *Athelia* sp. on hibiscus. (Courtesy J. Uchida)

80. Wilting of impatiens caused by Pythium root rot.

81. Healthy impatiens seedling (left) and seedling with root rot caused by *Pythium* sp. (right).

82. Petiole infection of African violet caused by *Pythium* sp.

83. Black leg of geranium caused by *Pythium* infection. (Courtesy E. M. Dutky)

84. Black leg of hybrid geranium seedlings caused by *Pythium* sp.

85. Vascular discoloration in a New Guinea impatiens stem caused by *Pythium irregulare.*

86. Combined effects of *Pythium ultimum* and fungus gnat larvae on geranium.

87. Root and stem rot of cineraria caused by *Phytophthora cryptogea.* (Copyright J. Fletcher)

88. Crown rot of African violet caused by *Phytophthora parasitica.*

89. Crown rot of gloxinia caused by *Phytophthora parasitica.* (Courtesy S. H. Kim)

90. Holiday cactus with crown rot caused by *Phytophthora parasitica.*

91. Discoloration and black streak on poinsettia stem caused by *Phytophthora parasitica.*

92. Internal discoloration of poinsettia stem caused by *Phytophthora parasitica.*

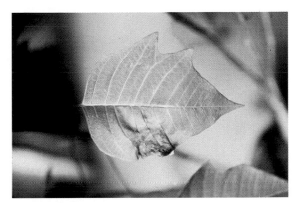

93. Poinsettia leaf lesion caused by *Phytophthora nicotianae.* (Courtesy J. Uchida)

94. Healthy (right) and *Phytophthora nicotianae*-infected poinsettia bracts. (Courtesy J. Uchida)

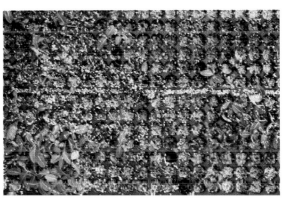

95. Damping-off of vinca (*Catharanthus roseus*) caused by *Phytophthora parasitica.*

96. Wilting of vinca infected with *Phytophthora nicotianae.* (Courtesy J. Uchida)

97. Gardenia plants infected with *Phytophthora parasitica.* (Courtesy A. R. Chase)

98. Gardenia plant with healthy root system (left) and three plants with increasingly severe levels of root rot caused by *Phytophthora parasitica.*

99. Canker on fuchsia caused by *Phytophthora parasitica.*

100. Phytophthora crown rot of African daisy (gerbera). (Courtesy J. A. Matteoni)

101. Downy mildew symptoms and signs on a *Ranunculus* sp. (Courtesy L. Pierce)

102. Chlorotic patches caused by downy mildew on the upper surface of a hardy geranium leaf.

103. White patches of downy mildew sporulation on the underside of a hardy geranium leaf.

104. Powdery mildew on cineraria.

105. Powdery mildew on kalanchoe. (Courtesy W. J. Manning)

107. Purpling and chlorosis of verbena leaves infected with powdery mildew.

106. Necrotic patches on kalanchoe leaves caused by powdery mildew.

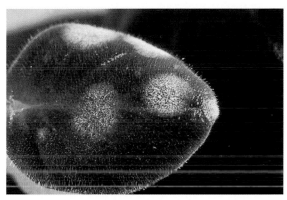

108. Powdery mildew on African violet. (Courtesy A. R. Chase)

109. Dark, greasy spots caused by powdery mildew on the underside of a begonia leaf.

110. Powdery mildew on begonia.

111. Necrosis of the upper leaf surfaces of begonia infected with powdery mildew.

112. Powdery mildew on a hiemalis (Rieger) begonia flower. (Courtesy R. K. Jones)

113. Necrotic patches on begonia leaves caused by powdery mildew.

114. Powdery mildew on dahlia. (Courtesy M. Shurtleff)

115. Powdery mildew on African daisy (gerbera). (Courtesy A. R. Chase)

116. Necrotic patches on a hydrangea leaf caused by powdery mildew. (Courtesy E. M. Dutky)

117. Powdery mildew on poinsettia.

118. White colonies of powdery mildew on the under surface of a poinsettia leaf.

119. Chlorotic spots caused by powdery mildew on the upper surface of the poinsettia leaf shown in Plate 118.

120. Chlorotic spots on the upper surface (left) and rust pustules on lower surface (right) of florist's geranium leaves caused by *Puccinia pelargonii-zonalis*.

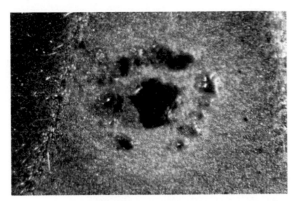

121. Secondary ring of uredia of geranium rust around the initial uredium on the underside of a geranium leaf.

122. Chlorotic spots on the upper surface of geranium leaves with geranium rust. (Courtesy M. F. Heimann)

123. Purple-rimmed rust lesion on the upper surface of a fuchsia leaf.

124. Rust sporulation on the underside of a fuchsia leaf.

125. Fusarium wilt of lisianthus (*Eustoma grandiflorum*). (Courtesy B. Løschenkohl)

126. Tip necrosis of lisianthus with Fusarium wilt. (Courtesy B. Løschenkohl)

127. Root rot of a young cyclamen caused by *Fusarium oxysporum* f. sp. *cyclaminis*.

128. Marginal chlorosis of cyclamen leaves on a plant with Fusarium wilt. (Courtesy J. A. Matteoni)

129. Chlorosis of cyclamen infected by both *Fusarium oxysporum* f. sp. *cyclaminis* and a soft rot *Erwinia* sp. (Courtesy A. R. Chase)

130. Discoloration in cyclamen corm (right) infected with *Fusarium oxysporum* f. sp. *cyclaminis*. (Courtesy J. A. Matteoni)

131. Chlorosis and wilting of lower leaves of geranium with Verticillium wilt.

132. Bacterial blight of poinsettia, caused by *Curtobacterium flaccumfaciens*. (Courtesy L. Pierce)

133. Dieback of poinsettia caused by *Curtobacterium flaccumfaciens*. (Courtesy A. W. Engelhard)

134. Bacterial soft rot of a cyclamen corm caused by *Erwinia* sp.

135. Erwinia soft rot of poinsettia.

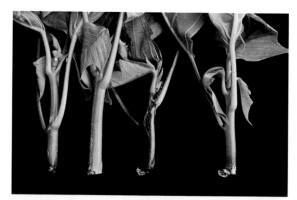

136. Soft rot of poinsettia cuttings caused by *Erwinia chrysanthemi*.

137. Erwinia soft rot of holiday cactus. (Courtesy A. R. Chase)

138. Leaf spot of *Bougainvillea* caused by *Pseudomonas andropogonis.* (Courtesy A. R. Chase)

139. Irregular, black leaf spots caused by *Pseudomonas cichorii* on florist's geranium. (Courtesy A. W. Engelhard)

140. Leaf spot of African daisy (gerbera) caused by *Pseudomonas cichorii.* (Courtesy A. R. Chase)

141. Leaf spot of hibiscus caused by *Pseudomonas cichorii.* (Courtesy A. R. Chase)

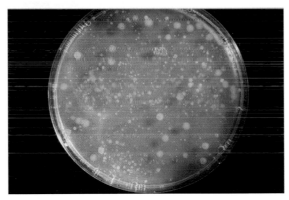

142. *Pseudomonas cichorii* on King's medium B without UV light. (Courtesy J. A. Matteoni)

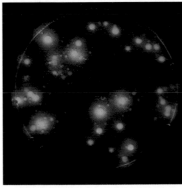

143. *Pseudomonas cichorii* on King's medium B with UV light. (Courtesy J. A. Matteoni)

144. Southern wilt of florist's geranium, caused by *Pseudomonas solanacearum.* (Courtesy R. K. Jones)

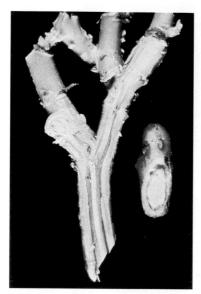

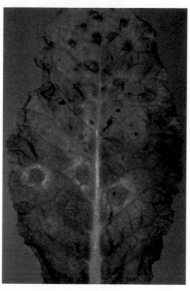

145. Internal vascular discoloration in florist's geranium infected by *Pseudomonas solanacearum*. (Courtesy R. K. Jones)

146. Spots on primrose caused by *Pseudomonas syringae* pv. *primulae*. (Courtesy C. Krass)

147. Necrotic spots and leaf distortion of hibiscus caused by *Pseudomonas syringae* infection. (Courtesy A. R. Chase)

148. Spotting and leaf-edge necrosis of impatiens caused by *Pseudomonas syringae*. (Courtesy A. R. Chase)

149. Water-soaked lesion on New Guinea impatiens caused by *Pseudomonas syringae*. (Courtesy A. R. Chase)

150. Hybrid seedling geranium leaves with bacterial leaf spots caused by *Pseudomonas syringae*.

151. Necrosis and chlorosis of a florist's geranium resulting from inoculation with *Pseudomonas syringae*.

152. Stunted basal shoots typical of bacterial fasciation of florist's geranium, caused by *Rhodococcus fascians*.

153. Bacterial leaf spot of hibiscus caused by *Xanthomonas campestris*. (Courtesy A. R. Chase)

154. Chlorosis and wedge-shaped area of necrosis in a florist's geranium (*Pelargonium* x *hortorum*) leaf caused by systemic infection with *Xanthomonas campestris* pv. *pelargonii*.

155. Small, round spots and wedges of necrosis in florist's geranium leaves infected by *Xanthomonas campestris* pv. *pelargonii*.

156. Leaf spots on ivy geranium (*Pelargonium peltatum*) caused by *Xanthomonas campestris* pv. *pelargonii*.

157. Purple leaf spots on wild crane's-bill (*Geranium* sp.) caused by *Xanthomonas campestris* pv. *pelargonii*. (Courtesy S. H. Kim)

158. Wilting of lower leaves of a florist's geranium systemically infected with *Xanthomonas campestris* pv. *pelargonii*.

159. Chlorosis and necrosis of an ivy geranium leaf systemically infected with *Xanthomonas campestris* pv. *pelargonii.*

160. Necrosis of lower leaves of an ivy geranium systemically infected with *Xanthomonas campestris* pv. *pelargonii.*

161. Necrotic spotting on regal geranium (*Pelargonium x domesticum*) caused by *Xanthomonas campestris* pv. *pelargonii.* (Courtesy S. H. Kim)

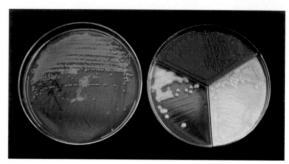

162. Characteristics of *Xanthomonas campestris* pv. *pelargonii* on various culture media. Left, King's medium B: pale yellow colonies; right, SX medium (purple): sparse or no growth; YDC medium (white): highly mucoid, yellow colonies; SPA medium (highly translucent): mucoid, pale yellow colonies.

163. Discrete spots and chlorosis of nonstop begonia caused by *Xanthomonas campestris* pv. *begoniae.*

164. Wedge-shaped necrosis in hiemalis (Rieger) begonia leaves infected by *Xanthomonas campestris* pv. *begoniae.*

165. Spotting on poinsettia leaves caused by *Xanthomonas campestris* pv. *poinsettiicola.* (Courtesy A. W. Engelhard)

166. Necrotic spotting on ornamental pepper caused by *Xanthomonas campestris* pv. *vesicatoria.*

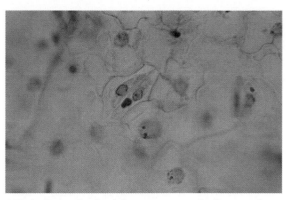

167. A partially virescent flower head in a hydrangea infected with a phytoplasma (previously referred to as a mycoplasmalike organism [MLO]).

168. Inclusion bodies of tomato spotted wilt tospovirus. (Courtesy G. W. Simone)

169. Local lesions caused by hydrangea ringspot potexvirus (left) and tomato spotted wilt tospovirus (right) on the indicator plant *Gomphrena globosa*. (Courtesy J. Pscheidt)

170. Leaf strapping and distortion on New Guinea impatiens caused by cucumber mosaic cucumovirus (CMV). (Courtesy G. Grüber)

171. Chlorosis and green island patterns in hydrangea leaves affected by hydrangea ringspot potexvirus. (Courtesy W. C. Stienstra)

172. Leaf chlorosis and green islands caused by kalanchoe mosaic potyvirus infection of kalanchoe. (Courtesy K. Bech)

173. Chlorotic ring spots caused by pelargonium flower break carmovirus on the upper leaf surface of florist's geranium (*Pelargonium* x *hortorum*). (Courtesy I. Bouwen)

174. Chlorotic ring spots caused by pelargonium flower break carmovirus on the lower leaf surface of florist's geranium (*Pelargonium* x *hortorum*). (Courtesy I. Bouwen)

175. Flower symptoms in florist's geranium caused by pelargonium flower break carmovirus. Top row: healthy; middle row: weak symptoms; bottom row: severe symptoms. Columns 1–4 are tetraploid cultivars, and column 5 is a diploid. (Courtesy W. Elsner)

176. Chlorotic rings and blotches on poinsettia caused by poinsettia mosaic virus. (Courtesy M. F. Heimann)

177. Stunting and leaf distortion in New Guinea impatiens infected with tobacco mosaic tombamovirus. (Courtesy R. A. Welliver)

178. Chlorotic patches on anemone leaves caused by impatiens necrotic spot tospovirus (INSV).

179. Browallia leaf showing necrosis at the point of petiole attachment caused by INSV.

180. Ornamental pepper infected with tomato spotted wilt tospovirus (TSWV) or INSV. (Courtesy W. R. Allen, Agriculture Canada)

181. Chlorosis and reddening of African daisy (gerbera) foliage infected with TSWV or INSV. (Courtesy W. R. Allen, Agriculture Canada)

182. Distortion and flecking of hibiscus leaves infected with INSV. (Courtesy R. A. Welliver)

183. Concentric, chlorotic rings in leaf of kalanchoe infected with INSV. (Courtesy R. A. Welliver)

184. Chlorotic mottling of leaves of lisianthus (*Eustoma grandiflorum*) infected with INSV. (Courtesy W. R. Allen, Agriculture Canada)

185. Chlorotic leaf areas in monkey-flower (mimulus) infected with TSWV. (Courtesy R. A. Welliver)

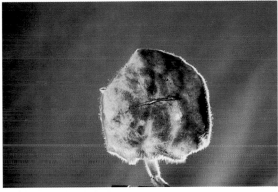

186. Ring spots in an African violet leaf caused by INSV. (Courtesy R. A. Welliver)

187. Chlorosis of cladophylls and a concentric ring spot on holiday cactus infected with INSV. (Courtesy R. A. Welliver)

188. Chlorotic leaf mottling of a begonia infected with INSV.

189. Necrotic veins in hiemalis (Rieger) begonia infected with INSV. (Courtesy R. K. Jones)

190. Chlorotic spotting in cineraria infected with INSV.

191. Black streaking of a petiole in cineraria infected with INSV. (Courtesy W. R. Allen, Agriculture Canada)

192. Leaf necrosis of calceolaria affected by INSV. (Courtesy W. R. Allen, Agriculture Canada)

193. Stunting of calceolaria affected by INSV.

194. Round, brown spots on a cyclamen leaf caused by INSV.

195. Leaf necrosis in a concentric ring pattern at the edge of a cyclamen leaf caused by INSV.

196. Necrotic line patterns in the foliage of a cyclamen affected by INSV.

197. Oak-leaf chlorotic line pattern in the foliage of dahlia infected with TSWV.

198. Chlorotic mottling of dahlia foliage infected with TSWV.

199. Stem blackening and tan cankers on side branches of exacum infected with INSV. (Courtesy R. K. Jones)

200. Leaf distortion and stunting of New Guinea impatiens infected with INSV.

201. Diffuse blackening in New Guinea impatiens leaves caused by INSV.

202. Purple and necrotic patches in leaves of New Guinea impatiens infected with INSV.

203. Ring spots in petals of New Guinea impatiens infected with INSV.

204. Stunting and black ring spots on double-flowered impatiens infected with INSV. (Courtesy J. Mishanec)

205. Ring spots on kalanchoe caused by INSV.

206. Chlorotic spotting of leaves of ivy geranium (*Pelargonium peltatum*) affected by TSWV. (Courtesy J. Fletcher)

207. Leaf stunting and necrosis in *Primula sinensis* affected by INSV.

208. Ranunculus with leaf necrosis and distortion caused by INSV. (Courtesy M. K. Hausbeck)

209. Healthy ranunculus (right) and a stunted plant infected with INSV. (Courtesy D. Karasevicz)

210. Effect of gloxinia plant age on severity of symptoms caused by INSV. Plants were exposed to INSV-infected western flower thrips at (clockwise from lower right) 4, 7, 11, 21, 28, and 35 days after transplanting. (Courtesy W. R. Allen, Agriculture Canada)

211. Necrotic line patterns in gloxinia infected with INSV.

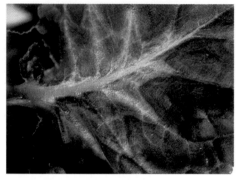

212. Chlorotic line pattern in gloxinia infected with INSV. (Courtesy W. R. Allen, Agriculture Canada)

213. Round, necrotic leaf spots on gloxinia caused by INSV.

214. Chlorotic ring spots on gloxinia caused by INSV. (Courtesy W. R. Allen, Agriculture Canada)

215. Symptoms of systemic INSV infection in the younger leaves of a gloxinia. (Courtesy W. R. Allen, Agriculture Canada)

216. Thrips feeding scars on a gloxinia leaf. The scar on the right shows the beginning of lesion development, indicating that transmission of INSV has occurred. (Courtesy W. R. Allen, Agriculture Canada)

217. Dark-rimmed spots at the sites of thrips feeding on a Calypso petunia leaf used as an indicator plant for the detection of tospovirus transmission in the greenhouse. (Courtesy L. S. Pundt)

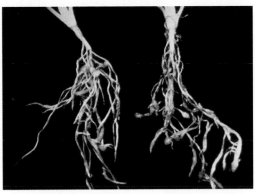

218. Swellings on African violet roots caused by root-knot nematodes. (Courtesy M. Gleason)

219. Chlorotic spotting on a lantana leaf caused by foliar nematode (*Aphelenchoides* sp.) infestation. (Courtesy G. Philley)

220. Reddening of a begonia leaf surface indicating foliar nematode (*Aphelenchoides* sp.) infestation. (Courtesy R. K. Jones)

221. Advanced symptoms of foliar nematode (*Aphelenchoides* sp.) infestation on begonia.

222. Interveinal discoloration on the underside of an African violet leaf caused by foliar nematode (*Aphelenchoides* sp.) infestation. (Courtesy R. K. Jones)

223. Cineraria leaves rolled in from the sides in response to exposure to ethylene.

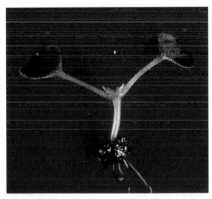

224. Death of the meristem tip of an impatiens seedling following exposure to ethylene.

225. Interveinal necrosis on a celosia plant caused by exposure to SO_2 from a greenhouse space heater.

226. Foliar chlorosis in gardenia caused by iron deficiency. (Courtesy A. R. Chase)

227. Molybdenum deficiency in poinsettia, causing leaf edge chlorosis and scorch. (Courtesy D. Cox)

228. Leaf edge necrosis in florist's geranium caused by an excessive level of soluble salts in the growing medium.

229. Wilting of florist's geranium caused by root injury from an excessive level of soluble salts in the growing mix.

230. Defoliation of impatiens resulting from excessive soluble salts.

231. Leaf-edge chlorosis in African violet caused by ammonium toxicity.

232. Leaf-edge necrosis and foliar flecking of florist's geranium caused by excessive levels of iron and manganese in the tissue. (Courtesy D. Cox)

233. Healthy florist's geranium (left) and one exhibiting advanced symptoms of iron and manganese toxicity.

234. Edema on the underside of a florist's geranium leaf. (Courtesy M. F. Heimann)

235. Injury to poinsettia leaves from application of a high rate of insecticidal soap.

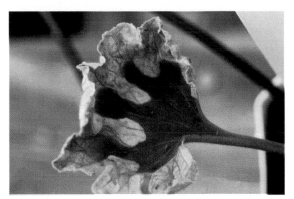

236. Temporary leaf-edge chlorosis caused by standard growth regulator application to a sensitive geranium cultivar.

237. Injury to geranium caused by uptake of 40 ppm of the herbicide atrazine. (Courtesy A. Senesac)

238. Injury to poinsettia caused by the herbicide 2,4-D. (Courtesy M. F. Heimann)

239. Cold water injury to foliage of African violet. (Courtesy L. Pierce)

240. Chlorine bleach uptake by poinsettia roots may result in black stem discoloration that takes the form of concentric, dark rings.

241. Twospotted spider mite feeding injury to an ivy geranium (*Pelargonium peltatum*) leaf.

242. White patches on *Primula* flower petals caused by thrips feeding. (Courtesy J. P. Sanderson)

243. Immature (larval) western flower thrips. (Courtesy J. P. Sanderson)

244. Broad mite feeding injury to begonias appears as bronze interveinal patches on leaf undersides.

245. Purplish discoloration along the veins of African daisy (gerbera) leaves infested with broad mites.

246. Leaf edges may curl under on New Guinea impatiens infested with broad mites.

247. Poinsettia cutting that failed to root because of feeding injury from fungus gnat larvae.

248. Sooty mold and cast skins of aphids on a hibiscus leaf. (Courtesy J. P. Sanderson)

TABLE 14. Rust Fungi that Affect Flowering Potted Plants

Potted Plant Host[a]	Other Host	Pathogen	Stages[b]
Anemone coronaria	*Prunus* spp.	*Tranzschelia pruni-spinosae* var. *discolor*	0, I
Aquilegia sp. (columbine)	Several genera in Poaceae	*Puccinia recondita*	0, I
Begonia × *hiemalis* (Rieger begonia)	?	*Pucciniastrum boehmeriae*	II*
Catharanthus roseus (vinca)	Autoecious	*Puccinia vincae*	0, I, II*, III
Euphorbia pulcherrima (poinsettia)	Autoecious	*Melampsora heliscopae*	0, I, II*, III
	Autoecious	*Uromyces euphorbiae*	II*, III
	Autoecious	*U. proeminens* var. *poinsettiae*	II*, III
Fuchsia × *hybrida*	Abies spp.	*Pucciniastrum epilobii*	II*
Lantana camara	Autoecious	*Puccinia lantanae*	III*
Pelargonium spp. (geranium)	Autoecious	*Puccinia pelargonii-zonalis*	II*, III
Pericallis × *hybrida* (cineraria)	?	*Coleosporium tussilaginis*	II*, III
Primula vulgaris (primrose)	?	*Puccinia primulae*	I, II*, III
	?	*Puccinia aristidae*	I
	Autoecious	*U. apiosporus*	
Ranunculus sp.	Several genera in Poaceae	*P. recondita*	0, I
	Autoecious	*Puccinia lagenophorae*	I, III

[a] In many cases the flower crop is the alternate host, i.e., the host not bearing the telia.
[b] 0 = Pycnial; I = aecial; II = uredial; and III = telial. * = Repeating stage occurs on potted plant host.

beneath those on the upper leaf surface and gradually enlarge into chlorotic, blisterlike pustules (Plate 120). Within 10–14 days after infection, these pustules break open, exposing the rusty brown urediospores. Additional uredia often form in a circular fashion around the original uredium (Plate 121). The yellow spots on the upper leaf surface and rust-colored spores on the lower surface distinguish geranium rust from other leaf spot diseases (Plates 121 and 122). Severely infected leaves become chlorotic and defoliate, rendering the crop unmarketable. The rust pustules (uredia) form mostly on the lower leaf surface but may also occur on the upper leaf surface, stems, petioles, and stipules.

Causal Organism

Puccinia pelargonii-zonalis Doidge is an autoecious rust with only uredial and telial stages. Urediospores (18–29 × 17–25 μm) are light to yellow brown and broadly ovate or subglobose (Fig. 38). These spores are finely echinulate and have two small, conspicuous, equatorial germ pores within the thin spore walls. The pale brown, two-celled, stalked teliospores are rarely observed on geraniums but when present can be mixed with urediospores in the uredia or formed separately in telia. The teliospores (36–57 × 19–26 μm) are ellipsoid or clavate, rounded at the apex, and slightly constricted at the septum; germ pores are located near the apex and below the septum.

Host Range and Epidemiology

Geranium rust has been reported on *Pelargonium* × *hortorum* L. H. Bailey (florist's geranium), *P. zonale* (L.) L'Hér., *P. inquinans* (L.) L'Her. ex Ait., *P. endlicherianum* Fenzl., *P. quercifolium* L'Hér., *P.* × *salmoneum* R. A. Dyer, and *P. elongatum* (Cav.) Salisb. (syn. *P. tabulare* L'Hér.). Not all *Pelargonium* species are susceptible, and considerable variation in susceptibility occurs among cultivars of *P.* × *hortorum*. *P.* × *domesticum* L. H. Bailey (regal geranium) is not a host for geranium rust, and some cultivars of *P. peltatum* (L.) L'Hér. (ivy geranium) are resistant. The disease has not been reported on *Geranium* spp.

Urediospores of *P. pelargonii-zonalis* require the presence of free water (for approximately 3 hr) and temperatures of 16–21°C for optimum germination. Disease increases most rapidly at 21°C. Urediospores remain infective in the greenhouse for at least 3 months and possibly longer.

Management

Since *P. pelargonii-zonalis* is an autoecious rust, the life cycle cannot be broken by elimination of an alternate host. The crop should be monitored carefully for symptoms, and

leaves with yellow spots should be turned over to check for rust pustules on the undersides (Plate 120). Infected plants should be discarded, and incoming plants should be inspected for evidence of rust. Leaves should be kept as dry as possible to reduce the germination potential of urediospores. Biological control of geranium rust with bacterial antagonists has been successful under experimental conditions and may be adopted in the future. The fungicides chlorothalonil, triadimefon, mancozeb, oxycarboxin, and sulfur are registered in the United States for the control of geranium rust.

Selected References

Harwood, C. A., and Raabe, R. D. 1979. The disease cycle and control of geranium rust. Phytopathology 69:923-927.

Kaufman, P. 1980. Control of geranium rust with foliar sprays. Fungic. Nematicide Tests 35:136-137.

Rytter, J. L., Lukezic, F. L., Craig, R., and Moorman, G. W. 1989. Biological control of geranium rust by *Bacillus subtilis*. Phytopathology 79:367-370.

Fuchsia Rust

Fuchsia rust occurs throughout the United States and Europe on *Fuchsia* × *hybrida* Hort. ex Vilm., particularly during the winter months. The most serious losses occur during propagation; however, diseased plants at any stage of growth are unmarketable. Fuchsia may recover from this disease, but defoliation significantly weakens the plants and occasionally results in death.

Symptoms

Rust on fuchsia first appears as small, discolored areas on the lower leaf surfaces. On the upper surfaces of the leaves, large, circular areas of chlorosis develop (Plate 123). As the disease progresses, masses of yellow orange urediospores are produced on the leaf undersides (Plate 124). Eventually spores are produced on both sides of the leaves, and the circular spots become less distinct. Occasionally, the spores form along the veins. Severely infected leaves yellow and drop prematurely.

Causal Organism

Fuchsia rust is caused by *Pucciniastrum epilobii* G. Otth f. sp. *palustris* Gaum (*P. pustulatum* Dietel in Engl. & Prantl), a heteroecious rust. Only urediospores are known to develop on fuchsia. These urediospores (10–14 × 13–23 μm) are obovate or oval and finely echinulate. The telial stage has not been reported on fuchsia.

43

Host Range and Epidemiology

P. epilobii occurs on *Fuchsia*, *Epilobium* (fireweed), and *Abies* (fir). *Fuchsia* and *Epilobium* are in the family Onagraceae, and both host the uredial stage of the rust. Teliospores, which allow the fungus to survive the winter, form on infected fireweed during the autumn. During the spring, the teliospores germinate and produce basidiospores, which infect the needles of various species of fir. Sexual conjugation of the fungus occurs on the fir needles followed by the development and release of aeciospores. The aeciospores can infect fireweed, and probably fuchsia, but cannot reinfect fir. The fungus then produces urediospores on infected fireweed and fuchsia. The urediospores reinfect fireweed and fuchsia repeatedly but cannot reinfect fir. The fungus can reside in the greenhouse, spread, and reinfect fuchsia as long as fuchsia or fireweed is present.

Management

The initial source of spores to infect fuchsia may be either fireweed or fir. Stock kept outdoors during the summer may become infected by airborne spores. Stock plants or cuttings brought in from elsewhere may also be sources of infection.

Fireweed in the vicinity of the greenhouse should be eliminated, and fuchsias that have rust should be discarded or isolated from healthy fuchsias. Mancozeb and sulfur are registered in the United States for use as protectant treatments against fuchsia rust. Plants should be spaced to provide good air circulation. Wetting the foliage during watering should be avoided. During cold weather, the greenhouse should be heated and ventilated at sunset to prevent condensation. Diseased stock plants can be cut back to woody stems, thus reducing the inoculum in the greenhouse. When the pruned plants begin to develop new foliage, they will be free of rust, but a protectant fungicide should be applied.

Selected Reference

Strider, D. L., and Jones, R. K. 1978. Rust of fuchsia in North Carolina. Plant Dis. Rep. 62:745-746.

Anemone Rust

Conspicuous pycnia are evenly, but sparsely, distributed over the upper leaf surface of *Anemone coronaria* L. (anemone). Cup-shaped, cinnamon brown aecia soon form on the lower leaf surface. As the aecia mature, the peridium characteristically splits into a few spreading lobes.

Anemone rust is caused by *Tranzschelia pruni-spinosae* (Pers.:Pers.) Dietel var. *discolor* (Fuckel) Dunegan. *T. pruni-spinosae* and *T. discolor* have also been reported on *A. coronaria* and are regarded as synonymous with *T. p. discolor*. The cinnamon brown *T. p. discolor* aeciospores (15–23 × 18–26 μm) are globose to oblong globose and have finely verrucose walls. Anemone rust occurs on *A. coronaria* and *Prunus* spp. The uredial and telial stages of this heteroecious rust are found on *Prunus* spp.

Selected Reference

Dunegan, J. C. 1938. The rust of stone fruits. Phytopathology 28:411-426.

Vascular Wilt Diseases

Vascular Wilts Caused by *Fusarium oxysporum*

Different formae specialis of *Fusarium oxysporum* cause vascular wilt diseases on *Cyclamen persicum* Mill., *Dahlia*,

Eustoma grandiflorum (Raf.) Shinn. (lisianthus), *Gerbera jamesonii* H. Bolus ex Adlam (African daisy), *Hibiscus rosasinensis* L., and *Ranunculus asiaticus* L. (see also Fusarium Wilt of *Cyclamen*). Other strains of *F. oxysporum*, which are not host specific, cause stem and root infections on other flowering potted plant hosts (see Fusarium Root, Crown, and Stem Rots).

Symptoms

Plants infected with *F. oxysporum* strains that cause wilt diseases generally exhibit symptoms of chlorosis and water deficiency (Plates 125 and 126). Plants may be stunted, and leaves may be scorched (symptoms are more pronounced on the older foliage). Highly susceptible cultivars may be killed, while others may show more moderate symptoms. Severe symptom development is often observed after a period of high temperatures.

Causal Organism

F. oxysporum Schlechtend.:Fr. generally forms macroconidia, microconidia, and chlamydospores abundantly in culture. Macroconidia are slightly sickle shaped, delicate, and thin walled. The apical cell is attenuated, and the basal shell is foot shaped. Macroconidia have three to five septa and are produced at first on branched, lateral phialides and later typically within sporodochia. The oval or kidney-shaped microconidia are produced in false heads and are most often single celled. The microconidia of *F. oxysporum* are produced on monophialides shorter than those of *F. solani*. The intercalary or terminal chlamydospores of *F. oxysporum* are formed singly or in pairs. There is no known teleomorph.

Morphology in culture is variable. Colonies on potato-dextrose agar have white aerial mycelium and show a fibrous margin. Most strains gradually develop a purplish tint visible from the underside. Cultures look similar to those produced by *F. subglutinans* but may be distinguished by the presence of chlamydospores and the monophialidic conidiophores in *F. oxysporum*. Sporodochia, when present, are cream, tan, or orange. The fungus may mutate in culture to forms with reduced sporulation and more aerial mycelium or to forms that appear wet and yellow to orange, a result of the production of macroconidia in pionnotes.

Host Range and Epidemiology

F. oxysporum formae specialis have restricted host ranges often limited to one plant species. *F. oxysporum* f. sp. *eustomae*, for example, is known to be pathogenic only to lisianthus. Fusarium vascular wilt disease usually begins with the intercellular invasion of the root in the zone of elongation. The fungus then moves into the xylem vessels in the area of differentiation, where it releases conidia that travel in advance of the hyphae. Eventually the fungus grows out through the cortex to the plant surface, where sporulation occurs. Plant-to-plant spread in the greenhouse occurs when spores present on the surfaces of plants are physically moved by handling or splashing irrigation water.

High greenhouse temperatures are optimum for Fusarium wilt. Fusarium wilt of African daisy (gerbera) is most often seen during midsummer when temperatures are 23–25°C. Dramatic summer symptom development is also characteristic of cyclamen wilt.

Nutritional factors influence wilt diseases caused by *F. oxysporum*. High levels of ammonium nitrogen, especially in the presence of insufficient potassium, are conducive to disease. Nitrate nitrogen applied exclusively causes some disease suppression, whereas ammonium nitrogen encourages disease. Low calcium levels also enhance Fusarium wilt, and sodium or other competitive cations may also increase wilt because they lead to reduced calcium uptake.

Trace element nutrition has an important effect on Fusarium wilt. Boron deficiency increases symptom development. Liming sandy soil to pH 6.5–7.5 controls Fusarium wilt by limiting the availability of some micronutrients. *Fusarium* is sensitive to micronutrient deficiency in the soil, while the host plant can efficiently extract nutrients in the rhizosphere. Adding excessive iron, manganese, zinc, or ammonium nitrogen can counteract the benefits of liming.

Management

Liming soils and using nitrate nitrogen fertilizer have been effective for management of *F. oxysporum* on *Dendranthema* × *grandiflorum* Kitam. (chrysanthemum), *Callistephus chinensis* (L.) Nees (aster), *Gladiolus* × *hortulanus* L. (gladiolus), *Cucumis sativus* L. (cucumber), *Lycopersicon esculentum* Mill. (tomato), and *Citrullus lanatus* (Thunb.) Matsum. & Nakai (watermelon). Although drenches with benzimidazole fungicides provide some disease suppression, fungicides are generally not sufficiently effective against Fusarium wilt.

Selected References

Engelhard, A. W. 1975. Aster Fusarium wilt: Complete symptom control with an integrated fungicide-NO$_3$-pH control system. (Abstr.) Proc. Am. Phytopathol. Soc. 2:62.

Engelhard, A. W., and Woltz, S. S. 1973. Fusarium wilt of chrysanthemum: Complete control of symptoms with an integrated fungicide-lime-nitrate regime. Phytopathology 63:1256-1259.

Fantino, M. G, Pasini, C., and Contarini, M. R. 1985. Lotta contro la fusariosi vascolare del ranuncolo. Colture Protette, no. 8/9, pp. 63-66.

Garibaldi, A., Gullino, M. L., and Rapetti, S. 1983. Tracheofusariosi: Una nuova malattia del ranuncolo. Colture Protette, no. 12, pp. 23-24.

Gordon, W. L. 1965. Pathogenic strains of *Fusarium oxysporum*. Can. J. Bot. 43:1309-1318.

Fusarium Wilt of *Cyclamen*

Fusarium wilt is an important and prevalent vascular wilt disease of *Cyclamen persicum* Mill. first reported from Germany in 1930 and from the United States in 1949. The disease was not reported from the Netherlands until 1977, but it is now common throughout the United States and Europe.

Symptoms

Symptoms may appear at any stage of plant growth and are strongly dependent upon inoculum level and growing conditions. The leaves of young seedlings may turn yellow and wilt when the roots are rotted, even though the corms have not yet developed vascular discoloration (Plate 127). Aboveground symptom development in infected plants may be delayed until the final stages of crop production. Plants exhibit yellow patches and wilting in one leaf after another until the whole plant collapses (Plates 128 and 129). This is characteristic of Fusarium wilt of cyclamen and helps to differentiate the disease from cyclamen stunt, which is caused by *Ramularia cyclaminicola*. The most useful character for diagnosis is the purple, reddish brown, or nearly black discoloration of the vascular system within the corm (Plate 130), which remains firm to the touch unless it is invaded by soft rot bacteria. Root rot may also occur. The distinctive vascular discoloration and the integrity of the corm clearly separate this disease from bacterial soft rot caused by *Erwinia* spp.

Causal Organism

Fusarium wilt of cyclamen is caused by *Fusarium oxysporum* Schlechtend.:Fr. f. sp. *cyclaminis* Gerlach. *F. oxysporum* is described in the previous section. Sporodochia of *F. o. cyclaminis* form a frosty coating at the bases of infected petioles and pedicels. The fungus is readily cultured from discolored vascular tissue of the corm.

Host Range and Epidemiology

The fungus infects only *C. persicum;* an attempt to transmit the disease to *Primula* (in the same plant family) was unsuccessful. Although there are several published opinions to the contrary, many plant pathologists believe that *F. o. cyclaminis* is initially introduced into the greenhouse on seed and then persists as a saprophyte. Seed transmission has not been conclusively demonstrated, but the pathogen has been isolated from organic debris in seed packets. Plants with latent infections are often passed from propagator to finished-plant producer.

The pathogen is capable of growth over a wide temperature range (6–35°C). Its optimum temperature of 28°C partially explains why peak losses occur during summer production. The fungus also grows well at 18–20°C, which is considered optimal for cyclamen production.

Disease development is influenced both by temperature and inoculum level. Twenty-five percent of the cyclamen grown in *Fusarium*-contaminated soil at 23°C developed symptoms within 2 weeks of transplant, while 100% were symptomatic within 4 weeks. At 17°C, symptoms did not develop for 6 weeks; and at 15°C, infection was latent until 7 weeks after transplanting. Low inoculum level delayed symptom development, but levels as low as 10 spores per milliliter caused symptoms in 60% of the inoculated plants after an extended incubation period (17 weeks).

The availability of iron affects the competitiveness of *Fusarium* spp. in soil. Iron added as ethylenediamine-di(*o*-hydroxyphenylacetic acid) (EDDHA) chelate has led to less infection in cyclamen than iron as a diethylenetriaminepentaacetic acid (DTPA)-chelate.

In several experiments with ebb-and-flood irrigation systems, spread from diseased to healthy cyclamen on the same table was not observed. Introduction of microconidia into either the recirculating nutrient solution or the potting soil did not lead to increases in inoculum levels in the nutrient solution. A sharp reduction in the population of the pathogen in the nutrient solution was seen within 24 hr. This decrease was attributed to sedimentation of the spores within the storage tank during the 24-hr period between irrigations. Only very high inoculum levels added to the recirculating nutrient solution resulted in disease development in cyclamen. One study determined that a concentration of 10^5 *Fusarium* spores per 100 ml of irrigation solution was necessary for infection but that this concentration was not reached as long as diseased plants were removed when detected.

Management

Since there are no available cyclamen cultivars with resistance to Fusarium wilt and chemical controls are not entirely effective, sanitation and fumigation of soil-containing media have been recommended. Treatments with bioantagonists and biologically active growing media have shown some promise for the suppression of disease caused by *F. o. cyclaminis*. Growing media prepared with dark, decomposed sphagnum peat are very conducive to disease development. In tests comparing different media for suppression of Fusarium wilt, several alternative media showed less disease development than sphagnum peat. Composted hardwood and pine bark media have shown promise for Fusarium wilt control in cyclamen and carnation. This effect may be attributable partly to the utilization of ammonium as the bark decomposes and partly to the role of biocontrol agents. Soil containing nonpathogenic *F. solani* and *F. oxysporum* antagonists of *F. o. dianthi* was also suppressive to *F. o. cyclaminis*. Earthworm composts also inhibited the pathogen.

Increasing the pH of the growing medium above 6.0 by the addition of dolomitic lime inhibits disease development.

Nitrate nitrogen fertilizer suppresses the disease, while ammonium nitrogen enhances it.

Frequent monitoring of the crop and prompt removal of diseased plants from the greenhouse bench are important for reducing disease spread via splashing irrigation water or insect activity. Removing diseased plants promptly from an ebb-and-flood system is important for keeping down the concentration of inoculum in the recirculating solution. Contaminated soil must be steam pasteurized or fumigated. Control of fungus gnats and other greenhouse insects may be important for reducing the number of spores introduced from a resident saprophytic soil population of the pathogen and for reducing plant-to-plant spread. Benzimidazole fungicides applied as drenches have shown some effectiveness against *Fusarium*, but the level of control is insufficient to ensure crop health. Spray treatments with thiophanate-methyl fungicides, which are registered in the United States for use on cyclamen, may help to minimize disease spread when used in conjunction with the roguing out of symptomatic plants.

Selected References

Daughtrey, M., and Hoitink, H. 1988. Control Fusarium wilt of cyclamen. Greenhouse Grower 6:34-42.

Garibaldi, A. 1988. Research on substrates suppressive to *Fusarium oxysporum* and *Rhizoctonia solani*. Acta Hortic. 221:271-277.

Gerlach, W. 1954. Untersuchungen über die Welkekrankheit des Alpenveichens (Erreger: *Fusarium oxysporum* Schl. f. *cyclaminis* n. f.). Phytopathol. Z. 22:125-176.

Grouet, D. 1985. Vascular Fusarium disease of cyclamen. Phytoma 372:49-51.

Krebs, E.-K. 1985. Erst bei hoher Sporendichte birgt Fusarium Gefahr. (Only with high spore densities is there a Fusarium risk.) Gb + Gw 85:1791-1793.

Rattink, H. 1986. Some aspects of the etiology and epidemiology of Fusarium wilt on cyclamen. Med. Fac. Landbouwwet. Rijksuniv. Gent 51:617-624.

Rattink, H. 1990. Epidemiology of Fusarium wilt in cyclamen in an ebb and flow system. Neth. J. Plant Pathol. 96:171-177.

Tompkins, C. M., and Snyder, W. C. 1972. Cyclamen wilt in California and its control. Plant Dis. Rep. 56:493-497.

Verticillium Wilt

Verticillium wilt, caused by two species of *Verticillium*, affects a broad range of species, including many of those grown as flowering potted plants. During recent decades, the widespread adoption of soilless growing media has reduced the likelihood that the pathogen will be introduced during greenhouse culture. Culture-indexing for cutting-propagated geraniums has sharply reduced the incidence of Verticillium wilt in this crop, but it is still seen occasionally in greenhouse culture.

Symptoms

Symptoms of Verticillium wilt include manifestations of water stress: stunting, leaf scorch, chlorotic foliage, and wilting. Foliar symptoms generally start in the lower leaves and progress upward. Lower leaves on infected impatiens yellow and drop, and plants wilt. The plant may recover. A characteristic of Verticillium wilt is that the symptoms may develop on one side of the plant.

Vascular discoloration may be present but is not a dependable trait for diagnosis. In *Pelargonium × hortorum* L. H. Bailey (florist's geranium), for example, the discoloration is light brown and hard to distinguish. In *Impatiens* spp., in contrast, the xylem may be blackened so that the discoloration is visible through the cortex of the stem (Fig. 39). *Verticillium* spp. can have a significant effect on the productivity (flower size and yield) of cut flowers such as *Gerbera jamesonii* H. Bolus ex Adlam (African daisy).

On *P. × hortorum*, wilting of a few leaves followed by general chlorosis occurs (Plate 131). Yellow, V-shaped wedges may develop in the leaves, which are affected progressively and turn brown and dry before abscising. Infected plants grow slowly and appear stunted. In the landscape, dwarfing may occur in the absence of other symptoms. Eventually the stems and inflorescences may blacken. Infection may be latent for several months before symptom development. Symptoms on *P. × domesticum* (regal geranium) are similar to those on *P. × hortorum*, except that the leaves tend to yellow without wilting. Symptoms of Verticillium wilt on *P. × hortorum* are very similar to those caused by systemic invasion of *Xanthomonas campestris* pv. *pelargonii*, the cause of bacterial blight.

On Rieger begonias, *Begonia × hiemalis* Fotsch., the initial symptoms are stunting and slow growth. Leaves exhibit increasingly severe chlorosis before becoming necrotic. *Impatiens wallerana* Hook f. with Verticillium wilt show wilting, stunting, leaf drop, stem dieback, and black streaks in the vascular system. Plant death has been observed in the landscape where plants were grown under water-stress conditions but not in greenhouse trials when water was applied as needed.

Causal Organisms

Two cosmopolitan species, *V. albo-atrum* Reinke & Berthier and *V. dahliae* Kleb., cause vascular wilt of a number of potted plants. The species can be distinguished by the fact that *V. dahliae* produces microsclerotia in culture, whereas *V. albo-atrum* produces only a dark, resting mycelium with hyphae 3–7 µm in diameter after 10–15 days in agar culture (Fig. 40). *V. albo-atrum*, unlike *V. dahliae*, does not grow at 30°C. The bases of *V. albo-atrum* conidiophores on plant tissue may be brownish, whereas *V. dahliae* conidiophores are always hyaline.

Both species have verticillately branched conidiophores; there are two to four phialides per node for *V. albo-atrum* and three to four for *V. dahliae* (Fig. 41). Phialides are mainly 20–30 (but up to 50) × 1.4–3.2 µm for *V. albo-atrum* and 16–35 × 1–2.5 µm for *V. dahliae*. The conidia are produced singly at the tips of the phialides and are ellipsoidal to irregularly sub-

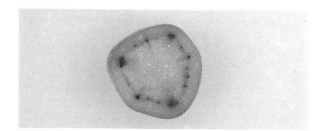

Fig. 39. Xylem discoloration of impatiens with Verticillium wilt.

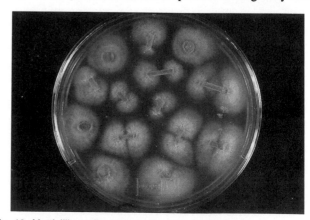

Fig. 40. *Verticillium albo-atrum* cultured from geranium.

Fig. 41. Verticillate conidiophores of *Verticillium* sp.

cylindrical, hyaline, and primarily one celled. Conidia of *V. albo-atrum* measure 3.5–10.5 (up to 12.5) × 2-4 µm and those of *V. dahliae* 2.5–8 × 1.4–3.2 µm.

Host Range and Epidemiology

Reported flower hosts of *V. albo-atrum* include *Abutilon, Aconitum, Antirrhinum, Aster, Begonia, Browallia, Calceolaria, Callistephus, Capsicum, Coreopsis, Dahlia, Delphinium, Dendranthema* (chrysanthemum), *Digitalis, Fuchsia, Gerbera, Impatiens, Matthiola, Paeonia, Pelargonium, Pericallis, Phlox, Portulaca, Rosa, Rudbeckia, Solanum,* and *Tagetes. V. dahliae* has been reported from *Abutilon, Begonia, Capsicum, Dianthus, Euphorbia, Impatiens, Rosa, Solanum,* and *Tagetes.* Host specialization is not known to occur.

Verticillium spp. are soilborne (hence the reduction of the importance of this disease with the widespread adoption of soilless media) and can survive in soil for years in the absence of a host.

Plants grown outdoors and then moved into the greenhouse (such as roses or tuberous dahlias) can be a source of contamination. The fungus may also be carried within symptomless cuttings. Culture-indexing of geraniums has greatly reduced the incidence of Verticillium wilt in that crop.

A temperature of at least 65°C is necessary during composting to eliminate *Verticillium* propagules. Thermal therapy of *Pelargonium* for *Verticillium* was found to be ineffective because the plant's heat sensitivity was greater than that of the fungus.

Greenhouse insect pests may play a role in crop contamination or dissemination of *Verticillium* spp. The fungus gnat, *Bradysia impatiens,* was shown to be a vector of *V. albo-atrum* during greenhouse culture of alfalfa. The insect introduced the fungus into feeding wounds made on the roots, stems, or leaves.

Management

Culture-indexed stock should be purchased when available, and symptomatic plants should be removed and destroyed. Insect pest populations that might serve as vectors should be reduced.

In general, fungicides have not been found to give adequate protection against *Verticillium* diseases. Drenches with benzimidazoles, such as thiophanate-methyl, have shown some benefit but have failed to provide complete control. Steam pasteurization is more effective and safer than fumigants for the treatment of soil to eliminate *Verticillium.*

Selected References

Coosemans, J. 1979. Partial control by systemic nematicides and/or fungicides of complex diseases caused by *Pratylenchus penetrans* (Cobb) Filipjev & Stekhoven, *Verticillium albo-atrum* Reinke & Berthold, and *Verticillium dahliae* Klebahn. Parasitica 35:63-72.

Grasso, S., and Pacetto, M. 1969. Una nueva tracheoverticilliosi (*Verticillium dahliae* Kleb.) della poinsettia (*Euphorbia pulcherrima* Willd.) in Sicilia. Tech. Agric. Catania 21:3-6

Hawksworth, D. L., and Talboys, P. W. 1970. *Verticillium albo-atrum.* Descriptions of Pathogenic Fungi and Bacteria, no. 255. Commonwealth Mycological Institute, Kew, England.

Hawksworth, D. L., and Talboys, P. W. 1970. *Verticillium dahliae.* Descriptions of Pathogenic Fungi and Bacteria, no. 256. Commonwealth Mycological Institute, Kew, England.

Kalb, D. W., and Millar, R. L. 1986. Dispersal of *Verticillium albo-atrum* by the fungus gnat (*Bradysia impatiens*). Plant Dis. 70:752-753.

McWhorter, R. P. 1962. Diverse symptoms in *Pelargonium* infected with *Verticillium.* Plant Dis. Rep. 46:349-353.

Nadakavukaren, M. J., and Horner, C. E. 1959. An alcohol agar medium selective for determining *Verticillium* microsclerotia in soil. Phytopathology 49:527-528.

Powell, N. T. 1971. Interactions of plant-parasitic nematodes with other disease causing agents. Pages 119-135 in: Plant Parasitic Nematodes, vol. 2. B. M. Zuckerman, W. F. Mai, and R. A. Rohde, eds. Academic Press, New York.

Taylor, N. J. 1993. First report of Verticillium wilt of *Impatiens wallerana* caused by *Verticillium dahliae* in the United States. Plant Dis. 77:429.

Vigodsky, H. 1969. Verticillium control in Gerbera. Plant Dis. Rep. 53:570-572.

Diseases Caused by Bacteria

Bacteria are microscopic, prokaryotic organisms that are involved in a diverse array of important natural processes, including degradation of organic matter, nitrogen fixation, and transformation and release of nutrients in soil, as well as pathogenesis of plants and animals. Bacteria cause crop losses of several hundred million dollars annually in the United States. Flowering potted plants are subject to a number of devastating bacterial diseases, most commonly caused by *Erwinia, Xanthomonas,* and *Pseudomonas* spp. *Agrobacterium, Curtobacterium,* and *Rhodococcus* spp. are less common problems but can result in economic losses.

Selected References

Bradbury, J. F. 1986. Guide to Plant Pathogenic Bacteria. C.A.B. International, Farnham Royal, England.

Dowson, W. J. 1957. Plant Diseases due to Bacteria. 2nd ed. Cambridge University Press, London.

Goto, M. 1992. Fundamentals of Bacterial Plant Pathology. Academic Press, New York.

Lelliott, R. A., and Stead, D. E. 1987. Methods for the Diagnosis of Bacterial Diseases of Plants. Blackwell Scientific Publications, Oxford.

Schaad, N. W., ed. 1988. Laboratory Guide for Identification of Plant Pathogenic Bacteria. 2nd ed. American Phytopathological Society, St. Paul, MN.

Crown Gall

Crown gall is caused by *Agrobacterium tumefaciens.* Because the pathogen is soilborne, the disease is particularly troublesome to certain fruit and nursery crops but only occasionally occurs on flowering potted plants. The bacterium is of considerable interest to science because of its similarities to oncogenic-type cancers of animals. *A. tumefaciens* transfers bacterial oncogenes to the plant, resulting in an overproduction of auxin and cytokinin, which in turn results in an unregulated and disorganized growth of plant cells to the extent that galls develop.

Symptoms

Galls usually form at the crown of the plant but occasionally occur on roots or on a stem or branch. The galls first appear as small masses of callus tissue but may enlarge to several centimeters in diameter. The galls are firm, and their surfaces are roughened and irregular (Fig. 42).

Causal Organism

A. tumefaciens (Smith & Townsend) Conn. is a gram-negative, aerobic, rod-shaped bacterium with peritrichous flagella. Three biovars have been proposed to account for differences in genotypic traits. Many isolates from soil or plants are not pathogenic and cannot be differentiated from pathogenic isolates on the basis of standard physiological or biochemical tests.

Fig. 42. Crown gall of geranium caused by *Agrobacterium tumefaciens.* (Courtesy Department of Plant Pathology, Cornell University)

Host Range and Epidemiology

The host range of *A. tumefaciens* is large. Angiosperms are the primary hosts, and about 600 species are known to be susceptible. The following flowering potted plants are known to be natural hosts or have been proven to be susceptible by inoculation: *Anemone* sp., *Begonia* Semperflorens-Cultorum (wax begonia) and Rex-Cultorum (rex begonia) hybrids, *Campanula* spp., *Capsicum annuum* L. (ornamental pepper), *Catharanthus roseus* (L.) G. Don (vinca), *Dahlia* hybrids, *Euphorbia pulcherrima* Willd. ex Klotzsch (poinsettia), *Fuchsia* × *hybrida* Hort. ex Vilm., *Hydrangea* sp., *Kalanchoe blossfeldiana* Poelln., *Lantana camara* L., *Pelargonium* × *domesticum* L. H. Bailey (regal geranium), *P.* × *hortorum* L. H. Bailey (florist's geranium), *P. peltatum* (L.) L'Hér. (ivy geranium), *Pericallis* × *hybrida* R. Nordenstam (cineraria), *Primula* sp., *Saintpaulia ionantha* Wendl. (African violet), and *Schizanthus* sp. (butterfly flower).

A. tumefaciens is a soilborne organism that enters plants through wounds that occur on the roots or at the crown just under the soil surface. Flowering potted plants are infrequently infected by *A. tumefaciens* unless they are brought into contact with field soil or are propagated from infected stock plants.

Management

Management practices employed for other soilborne diseases are appropriate for crown gall. When field soil is used in a potting medium, it should be pasteurized. Plants with symptoms should be discarded. Biological control of crown gall is possible with *A. radiobacter,* strain K84. However, crown gall occurs too infrequently to warrant preventive treatments.

Selected Reference

De Cleene, M., and De Ley, J. 1977. The host range of crown gall. Bot. Rev. 42:389-466.

Bacterial Blight of Poinsettia

Bacterial blight of *Euphorbia pulcherrima* Willd. ex Klotzsch (poinsettia) was first described in 1941 in New Jersey and subsequently reported in Florida, Maryland, New York, Pennsylvania, Virginia, and New Zealand. Disease outbreaks may result in extensive damage to the crop. Stems, leaves, and bracts may become blighted. In some cases, only branch cankers are noted, and disease incidence is often low in a particular crop.

Symptoms

The bacterium colonizes the intercellular region of the phloem and cortex, moving primarily vertically and giving rise to elongated, water-soaked to brown streaks visible on the surface of the stem (Plate 132). Radial colonization of cortical cells by the bacterium may also result in irregular water-soaked to brown cankers that girdle the stem. As the bacterium continues to move toward the growing point of the stem, the stem tip curves toward the colonized side. The terminal leaves may become deformed, and the subtending leaves may develop irregular brown spots and fall off (Plate 133). Severely cankered stems may develop longitudinal cracks. Amber droplets containing bacteria may be observed on affected leaves or stems.

Causal Organism

Bacterial blight is caused by *Curtobacterium flaccumfaciens* pv. *poinsettiae* (Starr & Pirone) Collins & Jones (previously *Corynebacterium flaccumfaciens* subsp. *poinsettiae*), a gram-positive, pleomorphic rod. Unlike most coryneform bacteria, *C. f. poinsettiae* is motile. The bacterium grows slowly in culture and is typically orange on nutrient broth-yeast extract agar, but shades of pink may also occur. Other characteristics are described in publications by Starr and Pirone and Vidaver and Davis. The snap bean pathogen, *C. f. flaccumfaciens* (Hedges) Collins & Jones, causes symptoms on poinsettia identical to those caused by *C. f. poinsettiae*.

Host Range and Epidemiology

Environmental sources of *C. f. poinsettiae* are not known. The bacterium has been cultured from asymptomatic, healthy-appearing, rooted poinsettia cuttings. In one survey, 10% of asymptomatic plants tested were found to harbor the bacterium. The bacterium can ooze from infected leaves and stems and be splashed by irrigation water to adjacent plants. Infection of leaves occurs through stomates. It is likely that pruning tools and hands can transmit the bacterium from plant to plant. Environmental conditions required for disease development have not been studied in detail, but warm conditions and high nitrogen levels have been reported to be associated with disease occurrence. Disease is not usually evident until the plants are close to finishing.

Management

Plants showing symptoms of this disease should be discarded, and stock plants suspected of harboring the bacterium should not be used for propagation. Because of the slow development of this disease, the location of the pathogen in the vascular system, and little if any plant-to-plant movement during the growing season, bactericidal sprays are not warranted.

Selected References

Davis, M. J. 1986. Taxonomy of plant-pathogenic coryneform bacteria. Annu. Rev. Phytopathol. 24:115-140.

McFadden, L. A., and Creager, D. B. 1960. Recent occurrence of a bacterial blight of poinsettia in Florida. Plant Dis. Rep. 44:568-571.

Salazar, M. P. 1983. Bacterial canker of poinsettia. M.S. thesis. Virginia Polytechnic Institute and State University, Blacksburg.

Starr, M. P., and Pirone, P. P. 1942. *Phytomonas poinsettiae* n. sp., the cause of a bacterial disease of poinsettia. Phytopathology 32:1076-1081.

Vidaver, A. K., and Davis, M. J. 1988. Coryneform plant pathogens. Pages 104-113 in: Laboratory Guide for Identification of Plant Pathogenic Bacteria. 2nd ed. N. W. Schaad, ed. American Phytopathological Society, St. Paul, MN.

Diseases Caused by Soft Rot *Erwinia* spp.

The genus *Erwinia* encompasses a heterogeneous group of bacteria. For convenience, the genus is commonly split into two groups: soft rotters and non-soft rotters. The soft rot *Erwinia* spp. cause important diseases of potted plants; the non-soft rotters do not. Only the soft rot *Erwinia* spp. will be covered here. They are fast-growing, opportunistic bacteria capable of causing serious losses within a few days.

Symptoms

Soft rot *Erwinia* spp. produce pectolytic enzymes, resulting in the dominant symptom, a soft, mushy rot of tissues. On a cutting, the base of the stem may be affected for several centimeters or more. Primulas develop a crown rot that may progress into the leaf petioles. Fleshy organs such as dahlia roots and cyclamen (*Cyclamen persicum* Mill.) corms develop a cheesy, often foul-smelling rot (Plate 134). Vascular infection may result in stunting and yellowing of plants and

collapse of stems when bacteria move into the cortex (Plate 135 and Fig. 43).

Soft rot of cuttings occurs more frequently than stem rot of rooted plants. Poinsettia (*Euphorbia pulcherrima* Willd. ex Klotzsch) cuttings are very susceptible to both *E. chrysanthemi* and *E. carotovora*. When conditions are optimum, symptoms develop within 24–48 hr. The stems may become rotted to a length of 5–10 cm (Plate 136). Plants that escape the cutting rot phase may later develop a range of symptoms. Stunting and vascular browning may occur when conditions are suboptimum for disease development. Vascular colonization and stem rot result in vascular browning, wilting, darkening, and water soaking of stems and leaves, often followed by complete collapse of the plant.

Both *E. chrysanthemi* and *E. carotovora* subsp. *carotovora* cause soft rot of *Kalanchoe blossfeldiana* Poelln. Symptoms may vary considerably, but wilting and collapse of the plant typically occur. Leaves may lose their luster and become flaccid and grayish, or chlorosis and necrosis may occur. Veins on infected leaves may turn chlorotic or dark brown to black. Petioles become dark brown to black as do terminals and areas of the stem. Vascular browning is evident.

Easter cactus (*Hatiora* spp.) is also susceptible. Symptoms include the loss of healthy green coloration in the cladophylls, progressing to necrosis, wilting, and collapse (Plate 137).

Causal Organisms

The species reported to cause soft rot of potted plants include *E. carotovora* subsp. *atroseptica* (van Hall) Dye, *E. carotovora* subsp. *carotovora* (Jones) Bergey et al, and *E. chrysanthemi* Burkholder, McFadden, & Dimock. *Erwinia* spp. are facultatively anaerobic, gram-negative rods that are motile by peritrichous flagella (except for one species). They are oxidase negative and catalase positive and produce acid from fructose, galactose, D-glucose, ß methyl glucoside, and sucrose. *E. c. carotovora* and *E. c. atroseptica* are closely related and can be distinguished chiefly by the inability of *E. c. atroseptica* to grow at 36°C. In addition, unlike *E. c. carotovora*, *E. c. atroseptica* produces reducing substances from sucrose and acid from α-methyl glucoside and utilizes maltose.

E. chrysanthemi differs from *E. carotovora* in its sensitivity to erythromycin, phosphatase activity, ability to produce gas from glucose, utilization of sodium malonate, and inability to produce acid from trehalose. *E. chrysanthemi* has been separated into six subgroups on the basis of phenotypic characteristics and the host from which each strain was originally cultured; however, there appears to be a weak relationship between physiological and biochemical characteristics and the host. Furthermore, there is no strong evidence for host specificity in *E. chrysanthemi*.

Host Range and Epidemiology

Flowering potted plants reported as hosts of soft rot *Erwinia* spp. are listed in Table 15. However, it is likely that many if

Fig. 43. Collapse of poinsettias infected with *Erwinia chrysanthemi*.

not all potted plants would be susceptible under some conditions to one or more species. *E. carotovora* and *E. chrysanthemi* have relatively broad host ranges, although each strain is often most virulent on the host from which it was isolated.

There has been little research on the epidemiology of soft rot diseases of specific potted plants; however, there are findings that are generally applicable. Soft rot bacteria may be associated with plants and plant debris, water, and soil or potting media. Surface and underground water have been shown to harbor *Erwinia* spp.; therefore, irrigation water should be considered a potential source as well as a means of dissemination of these bacteria. Insects may disseminate the bacteria and also provide an infection court by feeding. The inoculum source may change rapidly. For example, dahlia roots may be the initial source of *E. chrysanthemi;* but after a disease episode, the bacteria may become established in growing media or on tools and bench tops where routine activities lead to contamination continually. One of the enigmas of bacterial soft rot is that *Erwinia* spp. may be present throughout the crop cycle without causing disease.

Dissemination of soft rot bacteria occurs over long distances via the distribution of infested plant materials. In the greenhouse, water splash, contaminated tools, and handling of infected plants are long-recognized means of efficient dissemination.

Once they come in contact with the host, soft rot bacteria may remain as epiphytes without causing disease. The soft rot *Erwinia* spp. are opportunistic pathogens and require an impaired host and favorable environmental conditions to cause disease. Wounding during periods of favorable moisture and temperature facilitates infection and disease development. High nitrogen fertilization levels have been shown to increase resistance to *Erwinia* spp. in some plants; however, in poinsettia, high nitrogen levels significantly increased soft rot of cuttings in one trial.

Although the soft rot *Erwinia* spp. are similar in many ways, temperature alone can affect host range, symptom development, and virulence. *E. c. atroseptica* has an optimum temperature for growth of 27°C with a range of 3–35°C. *E. c. carotovora* has an optimum temperature of 28–30°C with a minimum of 6°C and a maximum of 37–42°C. *E. chrysanthemi* has the highest optimum temperature for growth of the three (34–37°C), and some strains can grow at temperatures above 45°C.

There is ample evidence that *E. carotovora* can survive in the soil and rhizosphere, the population being greatly affected by host root exudates. Even soilless media can become contaminated and harbor *Erwinia* spp. for several months in the absence of a host.

Management

Only plants thought to be free of soft rot *Erwinia* spp. should be vegetatively propagated. Stock plants should not be too soft or have excessive or deficient levels of nitrogen. Cuttings should be removed with a sharp knife. Cutting instruments should be disinfested with quaternary ammonium compounds regularly (sodium hypochlorite is corrosive to metal). Cuttings should be collected in a surface-disinfested container and transported to the propagation area as soon as possible to reduce stress from water loss. Bench surfaces should be thoroughly disinfested with sodium hypochlorite or a greenhouse disinfestant containing a quaternary ammonium compound. Bottom heat (21–22°C) should be applied to cuttings to facilitate rooting. This temperature should not exceed 24°C because high temperatures will stimulate bacterial growth.

Selected References

Dickey, R. S. 1979. *Erwinia chrysanthemi:* A comparative study of phenotypic properties of strains from several hosts and other *Erwinia* species. Phytopathology 69:324-329.

Dickey, R. S. 1981. *Erwinia chrysanthemi:* Reaction of eight plant species to strains from several hosts and to strains of other *Erwinia* species. Phytopathology 71:23-29.

Dickey, R. S., and Kelman, A. 1988. *Erwinia*. Pages 44-59 in: Laboratory Guide for Identification of Plant Pathogenic Bacteria. 2nd ed. N. W. Schaad, ed. American Phytopathological Society, St. Paul, MN.

Dinesen, I. G. 1979. A disease of *Kalanchoe blossfeldiana* caused by *Erwinia chrysanthemi*. Phytopathol. Z. 95:59-64.

Engelhard, A. W., McGuire, R. G., and Jones, J. B. 1986. *Erwinia carotovora* pv. *carotovora*, a pathogen of *Kalanchoë blossfeldiana*. Plant Dis. 70:575-577.

Haygood, R. A., Strider, D. L., and Echandi, E. 1982. Survival of *Erwinia chrysanthemi* in association with *Philodendron selloum*, other greenhouse ornamentals, and in potting media. Phytopathology 72:853-859.

Hoitink, H. A. J., and Daft, G. C. 1972. Bacterial stem rot of poinsettia, a new disease caused by *Erwinia carotovora* var. *chrysanthemi*. Plant Dis. Rep. 56:480-484.

Knauss, J. F., and Miller, J. W. 1974. Bacterial blight of *Saintpaulia ionantha* caused by *Erwinia chrysanthemi*. Phytopathology 64:1046-1047.

McCarter, S. M., Moody, E. H., and Waindle, M. L. 1988. A wilt and crown rot of *Primula* species caused by *Erwinia carotovora* subsp. *carotovora*. Plant Dis. 72:672-675.

McCarter-Zorner, N. J., Franc, G. D., Harrison, M. D., Michaud, J. E., Quinn, C. E., Sells, I. A., and Graham, D. C. 1984. Soft rot *Erwinia* bacteria in surface and underground waters in southern Scotland and in Colorado, United States. J. Appl. Bacteriol. 57:95-105.

Perombelon, M. C. M. 1982. The impaired host and soft rot bacteria. Pages 55-67 in: Phytopathogenic Prokaryotes, vol. 2. M. S. Mount and G. H. Lacy, eds. Academic Press, New York.

Perombelon, M. C. M., and Kelman, A. 1980. Ecology of the soft rot erwinias. Annu. Rev. Phytopathol. 18:361-387.

TABLE 15. Flowering Potted Plants Reported To Be Hosts of Soft Rot *Erwinia* spp.

Host	*E. carotovora* subsp. *atroseptica*	*E. carotovora* subsp. *carotovora*	*E. chrysanthemi*
Begonia sp.	−	+	+
Capsicum annuum (ornamental pepper)	−	+	+
Cyclamen persicum	−	−	+
Cyclamen sp.	−	+	+
Dahlia hybrids	−	+	+
Euphorbia pulcherrima (poinsettia)	−	+	+
Kalanchoe blossfeldiana	−	+	+
Pelargonium spp. (geranium)	+	+	+
Primula vulgaris, P. obconica (primrose), *Primula* Pruhonicensis hybrids (polyanthus)	−	+	−
Saintpaulia ionantha (African violet)	−	+	+
Schizanthus (butterfly flower)	+	+	−
Schlumbergera (holiday cactus)	−	+	−

Rogers, M. N. 1959. Decay of poinsettia cuttings by the soft rot bacterium, *Erwinia carotovora* (Jones) Holland. Plant Dis. Rep. 43:1236-1238.

Pseudomonas Leaf Spot
of *Bougainvillea*

Pseudomonas andropogonis was first described as causing bacterial stripe of *Zea mays* L. (corn), *Sorghum bicolor* (L.) Moench (sorghum), and *S. sudanense* (Piper) Stapf (Sudan grass). It is now known to have a wider host range that includes plants in a number of dicots. *Bougainvillea* was first described as a host in Zimbabwe in 1957. In 1986 it was reported on *Bougainvillea* in Australia, and in 1990 it was reported in the southeastern United States, including several locations in Florida.

Symptoms

Circular to irregular lesions with tan centers and reddish brown margins, with or without chlorotic halos, form on the foliage (Plate 138 and Fig. 44). A reddish brown necrosis of the leaf margin is common. Lesions may be delineated by veins. Puckering of leaves and defoliation may occur. Under conditions of high humidity and prolonged leaf wetness, foliar lesions may be black and bracts may become blighted.

Causal Organism

P. andropogonis (Smith) Stapp is a gram negative, rod shaped, obligately aerobic, nonfluorescent bacterium in the Pseudomonadaceae. *P. andropogonis* is unique in that it has a single polar, sheathed flagellum. It is positive for urease, catalase, ß-glucosidase, and phosphatase and utilizes acetate, malonate, and citrate. It is negative for oxidase, arginine dihydrolase, gelatin hydrolysis, levan production, and starch hydrolysis. *P. andropogonis* grows well in culture at 25–32°C but does not grow at 37°C. On nutrient agar, it grows more slowly than most pseudomonads. It is variable for tobacco hypersensitivity.

Host Range and Epidemiology

Strains isolated from *Bougainvillea* were reported to be pathogenic to *S. bicolor*, *Z. mays*, *Vicia sativa* L. (spring-vetch), *Dianthus caryophyllus* L. (carnation), *Gypsophila elegans* M. Bieb (baby's-breath), and *Limonium sinuatum* (L.) Mill. (statice). However, strains originating from these hosts are not all pathogenic to *Bougainvillea*. More extensive lists of hosts are given in Moffett et al and Bradbury.

Bougainvillea is particularly susceptible when plants are succulent and overhead irrigation is used, although in Florida the disease occurs commonly in the absence of these factors.

Temperatures most conducive to disease development are not known, but inoculations at temperatures of 18–30°C resulted in symptom development within 2 weeks.

Management

Plants with spotting should be discarded. Overhead irrigation should be avoided, or irrigation should take place during times when foliage will dry quickly.

Selected References

Bradbury, J. F. 1986. Guide to Plant Pathogenic Bacteria. C.A.B. International, Farnham Royal, England.

Moffett, M. L., Hayward, A. C., and Fahy, P. C. 1986. Five new hosts of *Pseudomonas andropogonis* occurring in eastern Australia: Host range and characterization of isolates. Plant Pathol. 35:34-43.

Walker, S. E. 1991. Bacterial leaf spot of bougainvillea in Florida caused by *Pseudomonas andropogonis*. Plant Dis. 75:968.

Leaf Spot Caused
by *Pseudomonas cichorii*

Symptoms

Pseudomonas cichorii causes lesions similar to those caused by *P. syringae*. On florist's geranium (*Pelargonium* × *hortorum* L. H. Bailey), the two bacteria cause leaf spots that are indistinguishable from each other (see Leaf Spots Caused by Pathovars of *Pseudomonas syringae*). Lesions on florist's geranium caused by *P. cichorii* may vary depending on environmental conditions. Plants subjected to rainfall tend to develop large (5–10 mm), irregularly shaped, dark brown to black lesions (Plate 139). In the greenhouse or when plants are subjected to infrequent wetting, lesions may be smaller with tan centers and dark margins. Yellowing invariably occurs. Peduncles and flower buds may also become infected (Fig. 45).

On African daisy (*Gerbera jamesonii* H. Bolus ex Adlam), tan to dark spots or lesions develop along veins (Plate 140). Hibiscus (*Hibiscus rosa-sinensis* L.) may develop large lesions with yellow halos bordered by purple margins (Plate 141).

Causal Organism

P. cichorii (Swingle) Stapp is a gram-negative rod that is fluorescent on King's medium B (Plates 142 and 143), oxidase positive, arginine dihydrolase negative, levan negative, and potato rot negative. It causes a hypersensitive reaction when infiltrated into tobacco leaves. Other characters useful for identification can be found in publications by Schaad, Lelliott, and Bradbury. *P. cichorii* can be differentiated from *P. syringae* by LOPAT tests (Table 16).

Host Range and Epidemiology

P. cichorii causes diseases of several ornamental and vegetable crops. *Aglaonema, Anthurium, Calathea, Caladium*

Fig. 44. Necrotic spots with yellow halos on *Bougainvillea* caused by *Pseudomonas andropogonis*. (Courtesy A. R. Chase)

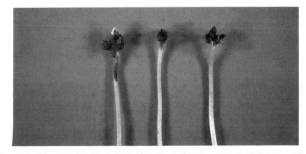

Fig. 45. Flower bud and peduncle infection on hybrid seedling geranium infected with *Pseudomonas cichorii*. (Courtesy A. W. Engelhard)

51

TABLE 16. LOPAT Scheme for Identification of Fluorescent *Pseudomonas* spp.

Test	*P. syringae*	*P. viridiflava*	*P. cichorii*	*P. agarici*	*P. marginalis*	*P. tolaasii*
Fluorescence	+	+	+	+	+	+
Oxidase	–	–	+	+	+	+
Arginine dihydrolase	–	–	–	–	+	+
Levan	+	–	–	–	+	–
Potato rot	–	+	–	–	+	–
Tobacco hypersensitivity	+	+	+	–	–	–

× *hortulanum* Birdsey, *Dieffenbachia, Dizygotheca, Epipremnum aureum* (Lindl. & André) Bunting (pothos), *Fatshedera, Fatsia japonica* (Thunb.) (Japanese aralia), *Hedera helix* L. (English ivy), *Monstera, Platycerium bifurcatum* (Cav.) C. Chr. (staghorn fern), *Philodendron, Polyscias,* and *Schefflera* are foliage plants reported to be hosts. Potted plant hosts include *Catharanthus roseus* (L.) G. Don (vinca), *Cyclamen persicum* Mill., *Dendranthema* × *grandiflorum* Kitam. (chrysanthemum), *G. jamesonii, H. rosa-sinensis, Jasminum, Pelargonium* spp. (geranium), and *Primula elatior* (L.) Hill. (primrose). Host specificity is not known to exist; isolates from chrysanthemum cause disease on *Pelargonium*. Of the plants listed above as hosts of *P. cichorii*, only *Pelargonium* and *Jasminum* are also hosts of *P. syringae* pv. *syringae* van Hall.

P. cichorii and *P. syringae* have similar life histories, but the optimum temperature for *P. cichorii* is 20–28°C. As is generally true for other bacterial diseases of foliage, periods of high humidity and leaf wetness are necessary for infection and disease development. At optimal temperature, moisture for a period of 48–72 hr during infection results in a high incidence of disease and lesion expansion. Chrysanthemums are known to carry epiphytic populations of *P. cichorii,* and this is probably the case with many other hosts. Thus, exchange of propagative material provides for long-distance distribution of *P. cichorii,* while irrigation and rain results in dispersal within the crop.

Management

Management practices outlined for *P. s. syringae* also apply to *P. cichorii*. Plants known to be carriers of *P. cichorii*, such as chrysanthemum, should be kept separate from other known hosts.

Selected References

Bradbury, J. F. 1986. Guide to Plant Pathogenic Bacteria. C.A.B. International, Farnham Royal, England.

Engelhard, A., Mellinger, H. C., Ploetz, R. C., and Miller, J. W. 1983. A leaf spot of the florists' geranium incited by *Pseudomonas cichorii*. Plant Dis. 67:541-544.

Jones, J. B., Raju, B. C., and Engelhard, A. W. 1984. Effects of temperature and leaf wetness on development of bacterial spot of geraniums and chrysanthemums incited by *Pseudomonas cichorii*. Plant Dis. 68:248-251.

Lelliott, R. A., and Stead, D. E. 1987. Methods for the Diagnosis of Bacterial Diseases of Plants. Blackwell Scientific Publications, Oxford.

Miller, J. W., and Knauss, J. F. 1973. Bacterial blight of *Gerbera jamesonii* incited by *Pseudomonas cichorii*. Plant Dis. Rep. 57:504-505.

Schaad, N. W., ed. 1988. Laboratory Guide for Identification of Plant Pathogenic Bacteria. 2nd ed. American Phytopathological Society, St. Paul, MN.

Southern Wilt

In the United States, *Pseudomonas solanacearum* causes a disease commonly known as southern bacterial wilt. A true soilborne pathogen, *P. solanacearum* typically enters the host through the root system and ultimately causes a vascular wilt.

A number of plants included in this compendium are susceptible; however, when soilless media is used, southern wilt is rarely a problem. The disease may occur when untreated field soil is used or when cuttings are taken from infected plants.

Symptoms

Wilting and yellowing of the lower leaves followed by necrosis are symptoms common to most crops colonized by *P. solanacearum* (Plate 144). Southern wilt almost always results in plant death. The vascular system is typically discolored. When cut stems are suspended in water, bacterial streaming is usually abundant. This bacterial streaming helps to differentiate southern bacterial wilt from wilts caused by fungi. In florist's geranium (*Pelargonium* × *hortorum* L. H. Bailey), lower leaves may wilt as soon as 2 weeks after infection. As the disease develops, leaves turn chlorotic then necrotic (Plate 144). The vascular system discolors, and the stem rots from the inside out, turning dark brown to black (Plate 145). Roots also become necrotic. Symptoms of southern wilt are similar to those caused by *Xanthomonas campestris* pv. *pelargonii*. However, *P. solanacearum* does not cause leaf spots, while *X. c. pelargonii* may or may not cause leaf spots.

Causal Organism

P. solanacearum (Smith) Smith is a gram-negative, rod-shaped, obligately aerobic, nonfluorescent bacterium in the Pseudomonadaceae. Some strains produce a diffusible brown pigment in culture. *P. solanacearum* is oxidase positive, accumulates poly-ß-hydroxybutyrate, and denitrifies. It does not form levan from sucrose, hydrolyze starch, or liquefy gelatin and is negative for arginine dihydrolase. Other characteristics can be found in publications by Harris, Hayward, and Lelliott.

Host Range and Epidemiology

P. solanacearum has a wide host range that includes at least 270 different plants in 44 families. Flowering potted plant genera susceptible to *P. solanacearum* include *Browallia, Capsicum, Catharanthus, Cyclamen, Dahlia, Fuchsia, Gerbera, Hydrangea, Impatiens, Lantana, Pelargonium, Pericallis,* and *Schizanthus*. A number of weed species are also hosts. Several biovars or races, which differ in host range, phenology, and response to environmental conditions, have been described. Race specificity has not been determined for most of the flowering potted plants listed.

Infection of plants generally occurs through the root system. High temperatures (30–35°C) and high soil moisture levels are generally conducive to disease development. Seed transmission is a possible but an unlikely means of infection. The bacterium has been shown to colonize foliage of *Capsicum* from infested seed. Water splash can disperse epiphytic populations or bacteria exuded from infected plants. Several weed hosts have been reported to be symptomless carriers of *P. solanacearum*.

Management

Southern wilt and other soilborne diseases can be avoided by using soilless growing media and disease-free propagative material. No chemicals are known to assist in southern wilt control.

Selected References

Harris, D. C. 1972. Intra-specific variation in *Pseudomonas solanacearum*. Pages 289-292 in: Proc. Int. Conf. Plant Pathog. Bact., 3rd.

Hayward, A. C. 1964. Characteristics of *Pseudomonas solanacearum*. J. Appl. Bacteriol. 27:265-277.

Hayward, A. C. 1991. Biology and epidemiology of bacterial wilt caused by *Pseudomonas solanacearum*. Annu. Rev. Phytopathol. 29:65-87.

Lelliott, R. A., and Stead, D. E. 1987. Methods for the Diagnosis of Bacterial Diseases of Plants. Blackwell Scientific Publications, Oxford.

Strider, D. L. 1982. Susceptibility of geraniums to *Pseudomonas solanacearum* and *Xanthomonas campestris* pv. *pelargonii*. Plant Dis. 66:59-60.

Strider, D. L., Jones, R. K., and Haygood, R. A. 1981. Southern bacterial wilt of geranium caused by *Pseudomonas solanacearum*. Plant Dis. 65:52-53.

Leaf Spots Caused by Pathovars of *Pseudomonas syringae*

Pseudomonas syringae van Hall was first described as a pathogen of lilac, *Syringa vulgaris* L., for which it was named. The bacterium has a wide host range that includes woody species, vegetable crops, grasses, and herbaceous ornamentals. To accommodate considerable differences in host specificity, *P. syringae* has been assigned many pathovar designations. Of all the pathovars, *P. s. syringae* has the widest reported host range and is the most important to flowering potted plants. *P. s. antirrhini* (Takimoto) Young, Dye, & Wilkie has been shown to be a pathogen of *Calceolaria* Herbeohybrida group by inoculation, and *P. s. primulae* (Ark & Gardner) Young, Dye, & Wilkie has been occasionally reported on *Primula* spp. (primrose) (Plate 146). A bacterial leaf spot of *Hibiscus rosa-sinensis* L. is caused by *P. s. hibisci* Jones et al.

Symptoms

Pathovars of *P. syringae* typically cause leaf spots characterized by water-soaked lesions, which may become dark brown to black, gray, or tan. Desiccation of infected tissue often results in a thin, papery lesion, which cracks as the leaf expands. When partially expanded leaves become infected, spotting and distortion of the expanded leaves occur (Plate 147). On *Impatiens wallerana* Hook. f., small spots with purple margins are typical, often originating at the leaf margin (Plate 148). New Guinea impatiens usually develop fairly large, water-soaked lesions, which turn dark brown (Plate 149). The florist's geranium (*Pelargonium* × *hortorum* L. H. Bailey) develops lesions (Plates 150 and 151) that are indistinguishable from those caused by *P. cichorii* (Swingle) Stapp. Small brown to black spots develop, which may coalesce into large, necrotic areas of the leaf. Yellowing of adjacent tissues occurs within a few days after lesions appear. Affected leaves die and remain attached to the plant without wilting. On *H. rosa-sinensis*, spots are irregularly shaped, dark brown lesions, 0.5–3 mm in diameter, with or without chlorotic halos (Plate 147). Blighting of foliage has also been reported.

Causal Organisms

P. syringae pathovars are gram-negative rods, fluorescent on King's medium B, oxidase negative, arginine dihydrolase negative, levan positive, and potato rot negative. They cause a hypersensitive reaction when infiltrated into tobacco leaves. Other characters useful for identification can be found in publications by Bradbury, Lelliott, and Schaad. *P. syringae* can be differentiated from other fluorescent pseudomonads by LOPAT tests (Table 16).

Host Range and Epidemiology

P. syringae pathovars have a very wide host range that includes *Dahlia* hybrids, *H. rosa-sinensis*, *I. wallerana*, New Guinea impatiens, and the florist's geranium among the potted plants. Sources of *P. syringae* in the potted plant industry have not been widely reported, but seed and plantlets are most likely the primary sources. Wild cherry trees (*Prunus* spp.), rye (*Secale cereale* L.), and various weeds have been reported to be epiphytic hosts. Lilacs may also be sources of the bacterium. In general, *P. syringae* pathovars are capable of causing disease at low temperatures (15–20°C). The optimum temperature for development of bacterial spot of *H. rosa-sinensis* is 18°C. Disease is exacerbated by high humidity and extended leaf wetness periods. *P. syringae* has been reported to be seedborne in several crop plants, including *Hibiscus*.

Management

Sanitation is an important disease-management principle and is particularly pertinent to bacterial problems. Plants that develop symptoms should be discarded. Workers should wash their hands after handling diseased plants or soil, and diseased plant debris should be removed from the growing area as promptly as possible. Bacteria are easily splashed from plant to plant by irrigation water. Splashing should be minimized and leaf wetness duration reduced as much as practical by irrigating early in the day or by subirrigating. Handling foliage when it is wet should be avoided. Nutrition may affect disease susceptibility; high levels of nitrogen have been reported to increase the susceptibility of *I. wallerana* to *P. s. syringae*. Copper sulfate pentahydrate and copper hydroxide are registered in the United States for control of *Pseudomonas* spp. Product labels list approved uses on flowering potted plants. Bactericides are only marginally effective in controlling bacterial diseases, but sanitation and environmental controls are extremely important.

Selected References

Bradbury, J. F. 1986. Guide to Plant Pathogenic Bacteria. C.A.B. International, Farnham Royal, England.

Cooksey, D. A., and Koike, S. T. 1990. A new foliar blight of *Impatiens* caused by *Pseudomonas syringae*. Plant Dis. 74:180-182.

Jones, J. B., Chase, A. R., Raju, B. C., and Miller, J. W. 1986. Bacterial leaf spot of *Hisbiscus rosa-sinensis* incited by *Pseudomonas syringae* pv. *hibisci*. Plant Dis. 70:441-443.

Lelliott, R. A., and Stead, D. E. 1987. Methods for the Diagnosis of Bacterial Diseases of Plants. Blackwell Scientific Publications, Oxford.

Nakada, N., and Takimoto, K. 1923. Bacterial blight of *Hibiscus*. Ann. Phytopathol. Soc. Jpn. 5:13-19.

Schaad, N. W., ed. 1988. Laboratory Guide for Identification of Plant Pathogenic Bacteria. 2nd ed. American Phytopathological Society, St. Paul, MN.

Wick, R. L., and Rane, K. K. 1987. *Pseudomonas syringae* leaf spot of *Pelargonium* × *hortorum*. (Abstr.) Phytopathology 77:1620.

Greasy Canker of Poinsettia

A canker of *Euphorbia pulcherrima* Willd. ex Klotzsch (poinsettia) caused by *Pseudomonas viridiflava* was described and named greasy canker in 1981. The disease was also reported in Florida, and the pathogen has been cultured from stems and foliage of diseased poinsettia in Massachusetts.

Symptoms

The first report of this disease described a canker and leaf spot as well as a bract and bud blight. Cankers may originate at pruning wounds on the stem. The cankered areas are greasy in appearance with no soft rot. They eventually turn light tan to brown and develop a papery texture as the cuticle lifts off the stem. Leaf spots are angular and randomly distributed (Fig. 46).

Causal Organism

P. viridiflava (Burkholder) Dowson is a gram-negative, fluorescent pseudomonad. It does not produce levan from sucrose and is negative for arginine dihydrolase and oxidase. It is potato rot positive and causes a hypersensitive reaction in tobacco (Table 16). It cannot utilize sucrose as a sole source of carbon, distinguishing it from other oxidase-negative fluorescent pseudomonads.

Host Range and Epidemiology

P. viridiflava was first described from green beans in 1927 and is now known to cause disease in a number of plants. An opportunistic pathogen, it is reported to be an epiphyte on some plants and a secondary invader. On several hosts, it causes a marginal leaf necrosis and necrotic spots on stems and fruits. Potted plant genera that have been reported as hosts include *Euphorbia, Hibiscus, Capsicum,* and *Hydrangea.*

There are conflicting reports of the optimum temperature for disease development in poinsettia. Suslow and McCain reported that severity was greater at 27–32°C than at 25°C. However, tests in Florida indicated that disease was severe at 10 and 15°C, mild at 27.7°C, and absent at 32.2°C. Because of this variability, it does not appear that temperature can be used to predict disease occurrence or severity for *P. viridiflava.*

Management

Control measures outlined for *P. syringae* are appropriate for *P. viridiflava.* Plants known to be infected should be discarded and not used for propagation. There are no bactericides known to be effective against diseases caused by *P. viridiflava.*

Selected References

Billing, E. 1970. *Pseudomonas viridiflava* (Burkholder, 1930; Clara 1934). J. Appl. Bacteriol. 33:492-500.

Engelhard, A. W., and Jones, J. B. 1989. *Pseudomonas viridiflava,* the cause of a stem canker of poinsettia plants. (Abstr.) Phytopathology 79:1168.

Suslow, T. V., and McCain, A. H. 1981. Greasy canker of poinsettia caused by *Pseudomonas viridiflava.* Plant Dis. 65:513-514.

Fig. 46. Necrotic spotting on poinsettia leaf caused by *Pseudomonas viridiflava.*

Bacterial Fasciation

Bacterial fasciation is not as prevalent in the flowering potted plant industry now as it was when field soil was used as the growing medium. The disease occurs sporadically and losses are usually minimal. It is most commonly observed on the florist's geranium, *Pelargonium × hortorum* L. H. Bailey.

Symptoms

Fasciation may take several forms. Cylindrical organs such as stems or peduncles become flattened and bandlike. Buds may proliferate, resulting in leafy, cauliflower-like galls, especially at the crown of the plant. Witches'-brooms may also occur. Necrosis of tissues usually does not occur. Stubby, stunted shoots are seen on florist's geranium (Plate 152).

Causal Organism

Rhodococcus fascians (Tilford) Goodfellow (previously *Corynebacterium fascians*) is a gram-positive, nonmotile, pleomorphic rod. It is cream to orange or yellow, depending on the isolate and the culture medium.

Host Range and Epidemiology

R. fascians has a wide host range that includes both monocots and dicots. Of the flowering potted plants, *Begonia* Tuberhybrida hybrids (nonstop begonia), *Dahlia* hybrids, *Impatiens wallerana* Hook. f., *Kalanchoe* sp., *P. × domesticum* L. H. Bailey (regal geranium), *P. × hortorum,* and *Schizanthus pinnatus* Ruiz & Pav. (butterfly flower) are hosts.

The ecology of *R. fascians* is not well known. The bacterium can be seedborne, and it is speculated that it is soilborne. It colonizes the cotyledonary buds as the plant emerges from the soil and is usually found on the surface of symptomatic meristematic tissues, where it produces growth regulators. The optimum temperature for growth is 24–27°C.

Management

Hot water treatment of seed effectively controls fasciation of nasturtium; however, seed is not an important source of this disease in flowering potted plants. The disease occurs too infrequently to warrant attention to preventive management practices. When fasciation occurs, plants should be discarded.

Selected References

Baker, K. F. 1950. Bacterial fasciation disease of ornamental plants in California. Plant Dis. Rep. 34:121-126.

Davis, M. J. 1986. Taxonomy of plant-pathogenic coryneform bacteria. Annu. Rev. Phytopathol. 24:115-140.

Diseases Caused by Pathovars of *Xanthomonas campestris*

Xanthomonas campestris (Pammel) Dowson causes diseases of a wide range of plants. In some cases, specific strains (designated pathovars) are restricted to narrow host ranges. For example, *X. c. begoniae* (Takimoto) Dye is pathogenic only to *Begonia,* and *X. c. pelargonii* (Brown) Dye is pathogenic only to *Geranium* and *Pelargonium.* Three pathovars have been described as pathogens of plants within the Euphorbiaceae: *X. c. euphorbiae* (Sabet, Ishag, & Khalil) Dye, *X. c. manihotis* (Berthet & Bondar) Dye, and *X. c. poinsettiicola* (Patel, Bhatt, & Kulkarni) Dye. All three have been reported to cause leaf spot on poinsettia, *Euphorbia pulcherrima* Willd. ex Klotzsch, yet each has a unique host range.

X. c. clerodendri (Patel, Kulkarni, & Dhande) Dye was described as a pathogen of *Clerodendrum phlomidis* L. f.; its pathogenicity to *C. thomsoniae* Balf. is unknown. *X. c. lantanae* (Srinivasan & Patel) Dye was described on outdoor plantings of *Lantana camara* L. in India. It was reported to occur only on *Lantana* species and has not as yet been reported in the potted plant industry. Bacterial leaf spots on *Hibiscus rosa-sinensis* L. may result from infection by either *Pseudomonas* spp. or *X. c. malvacearum* (Smith) Dye (Plate 153).

Selected References

Ark, P. A., and Barrett, J. T. 1946. A new bacterial leaf spot of greenhouse-grown gardenias. Phytopathology 36:865-868.

Dye, D. W. 1962. The inadequacy of the usual determinative tests for the identification of *Xanthomonas* spp. N.Z. J. Sci. 5:393-416.

Patel, M. K., Kulkarni, Y. S., and Dhande, G. W. 1952. Two new bacterial diseases of plants. Curr. Sci. 21:74-75.

Srinivasan, M. C., and Patel, M. K. 1957. Two new phytopathogenic bacteria on verbenaceous hosts. Curr. Sci. 26:90-91.

Bacterial Blight of Geranium

Bacterial blight is the most destructive disease of the florist's geranium, *Pelargonium* × *hortorum* L. H. Bailey. The disease has been referred to as bacterial stem rot, bacterial wilt, and bacterial leaf spot, but the more comprehensive term *blight* is preferable. Despite the long recognition of this disease and the widespread use of culture-indexing techniques (Fig. 47), bacterial blight remains a serious and frequent problem for geranium growers.

Symptoms

Symptoms vary depending on the cultivar and species affected, environmental conditions, and perhaps the strain of bacterium. When the bacterium is disseminated by splashing water, small, water-soaked spots develop, first visible on the leaf undersurface. After a few days, these leaf spots become apparent on the upper surface. Spots are tan to brown, round, and generally 2–3 mm in diameter. They are slightly sunken and have well-defined margins. Spots are typically followed by wedge-shaped areas of chlorosis and necrosis (Plates 154 and 155). Ivy geraniums (*P. peltatum* (L.) L'Hér.) and crane's-bills (*Geranium* spp.) develop lesions similar to those on florist's geraniums (Plates 156 and 157). Typically, the bacterium moves into the vascular system from the leaf spots and eventually causes wilt of the entire plant. Occasionally the bacterium does not spread from the leaf into the vascular system, and only the infected leaf dies. When zonal geraniums are systemically infected through roots or via latently infected

Fig. 47. Culture-indexing of geranium stock requires discarding plants if surface-sterilized sections from the base of a cutting show evidence of bacterial or fungal contamination when cultured in a broth medium. Cloudiness of the medium at left indicates bacterial growth.

cuttings, the first symptom is wilting of the lower leaves (Plates 154 and 158).

Because of the structure of ivy geranium leaves, systemic infection does not result in wilt. Ivy geranium foliage loses luster and develops symptoms suggestive of a nutrient deficiency or mite infestation (Plates 159 and 160). Regardless of the host, stem rot typically follows vascular infection when the bacterium moves from the vascular system into the pith and outward into the cortex. Systemically infected cuttings may or may not root. Those that do not root usually rot at the base. Leaf spot symptoms have been reported for the regal geranium, *P.* × *domesticum* L. H. Bailey (Plate 161).

Causal Organism

Bacterial blight of geranium is caused by *Xanthomonas campestris* pv. *pelargonii* (Brown) Dye. *X. campestris* (Pammel) Dowson is a gram-negative, rod-shaped, obligately aerobic, yellow-pigmented bacterium in the Pseudomonadaceae. It can be distinguished from *Pseudomonas* spp. by its production of xanthomonadins (yellow pigments), requirement for growth factors, and inability to utilize asparagine as a sole source of carbon and nitrogen. *X. c. pelargonii* has the characteristics of *X. campestris* but is restricted in its host range to the genera *Pelargonium* and *Geranium*. Cultural characteristics of *X. c. pelargonii* on agar media help to differentiate it from other pathovars (Plate 162). *X. c. pelargonii* is characteristically not as yellow as other xanthomonads. Occasionally, typical bright yellow xanthomonads other than *X. c. pelargonii* are cultured from geranium. *X. campestris* grows poorly on SX medium and utilizes starch slowly.

Host Range and Epidemiology

X. c. pelargonii is restricted to *Pelargonium* and *Geranium*. Cultivars of both *P. hortorum* (grown from cuttings or seed) and *P. peltatum* are susceptible to bacterial blight. No significant resistance among the commercially propagated cultivars has been noted. Crane's-bill is also a host. Some cultivars of regal or Martha Washington geranium, *P.* × *domesticum*, are quite resistant, but leaf spot has been observed (Plate 161). The sweet-scented geranium, *P. graveolens* L'Hér., is also a host and has intermediate susceptibility. *P. cordifolium* (Cav.) Curtis, *P. cucullatum* (L.) L'Hér. (maple-leafed geranium), *P. tomentosum* Jacq. (peppermint geranium), and *P.* × *scarboroviae* Sweet are resistant.

Leaf spot symptoms may occur within 7 days at 27°C or 21 days at 16°C. Temperatures below 10°C or above 32°C may suppress symptom development. Older, more woody plants appear to be less susceptible to systemic colonization through infection of the foliage. These plants may, however, harbor bacteria in the vascular system, resulting in infected daughter plants. Many instances have been observed of infected stock plants that do not show symptoms of disease at the time that cuttings are removed. Moisture and warmth supplied during rooting apparently trigger disease development.

X. c. pelargonii is capable of infection through the root system, although the bacterium does not survive in soil in the absence of infected host debris. It can survive for at least a year in undecomposed plant tissues. It can also survive on foliage of nonhost plants and has been reported to survive epiphytically on geranium leaves for several months without causing symptoms. *X. c. pelargonii* can also survive on wild or cultivated perennial *Geranium* spp. and overwinter as far north as Ithaca, New York. Overwintering in protected nursery crops of ornamental *Geranium* spp. or in mulched landscape plantings in cold temperate regions is particularly likely.

The cutting knife is a common means of disease transmission from one stock plant to another. Splashing irrigation water, removal of infected leaves, and dripping of water from infected ivy geraniums in hanging baskets to geraniums below are important means of dissemination. The greenhouse white-fly, *Trialeurodes vaporariorum* (Westwood), can transmit *X. c. pelargonii* from diseased to healthy geraniums. The foliage of plants other than geranium can harbor populations of *X. c. pelargonii*, but their importance as a source of this bacterium in greenhouse geranium production is not known.

Management

Strict adherence to good sanitation practices and a carefully administered culture-indexing program are the primary propagator's sole tools for supplying clean cutting stock. Secondary geranium propagators must purchase stock from reputable sources and rely on disease exclusion as much as possible. In the event of a bacterial blight outbreak, diseased plants must be rogued out promptly.

Hanging ivy geranium baskets should not be placed directly above geranium crops, and seed geraniums should be grown apart from vegetatively propagated cuttings. Geraniums from different propagators should be kept separate. Losses are more likely and may be much more extensive in greenhouse ranges in which geraniums of different types and from different sources are brought into close contact during production. No effective chemical control is known for this disease.

Selected References

Brown, N. A. 1923. Bacterial leafspot of geranium in the eastern United States. J. Agric. Res. 23:361-372.

Bugbee, W. M., and Anderson, N. A. 1963. Whitefly transmission of *Xanthomonas pelargonii* and histological examination of leafspots of *Pelargonium hortorum*. Phytopathology 53:177-178.

Burkholder, W. H. 1937. A bacterial leaf spot of geranium. Phytopathology 27:554-560.

Daughtrey, M., and Macksel, M. 1993. Nursery-cultivated *Geranium* spp. as a possible inoculum source for *Xanthomonas campestris* pv. *pelargonii* causing bacterial blight disease of a greenhouse *Pelargonium* crop. (Abstr.) Phytopathology 83:242.

Digat, B. 1978. Mise en evidence de la latence epiphylle du *Xanthomonas pelargonii* (Brown) Starr et Burkholder ches le *Pelargonium*. Ann. Phytopathol. 10:61-66.

Dimock, A. W. 1962. Obtaining pathogen-free stock by cultured cutting techniques. Phytopathology 52:1239-1241.

Dougherty, D. E., Powell, C. C., and Larsen, P. O. 1974. Epidemiology and control of bacterial leaf spot and stem rot of *Pelargonium hortorum*. Phytopathology 64:1081-1083.

Dunbar, K. B., and Stephens, C. T. 1992. Resistance in seedlings of the family Geraniaceae to bacterial blight caused by *Xanthomonas campestris* pv. *pelargonii*. Plant Dis. 76:693-695.

Kennedy, B. W., Pfleger, F. L., and Denny, R. 1987. Bacterial leaf and stem rot of geranium in Minnesota. Plant Dis. 71:821-823.

Kivilaan, A., and Scheffer, R. P. 1958. Factors affecting development of bacterial stem rot of *Pelargonium*. Phytopathology 48:185-191.

Knauss, J. F., and Tammen, J. 1967. Resistance of *Pelargonium* to *Xanthomonas pelargonii*. Phytopathology 57:1178-1181.

McPherson, G. M., and Preece, T. F. 1978. Bacterial blight of *Pelargonium*: Movement and symptom distribution of *Xanthomonas pelargonii* (Brown) Starr & Burkholder in *Pelargonium* × *hortorum* Bailey, following artificial inoculation. Pages 943-956 in: Proc. Int. Conf. Plant Pathog. Bact., 4th.

Munnecke, D. E. 1954. Bacterial stem rot and leaf spot of *Pelargonium*. Phytopathology 44:627-632.

Munnecke, D. E. 1956. Survival of *Xanthomonas pelargonii* in soil. Phytopathology 46:297-298.

Munnecke, D. E. 1956. Development and production of pathogen-free geranium propagative material. Pages 93-95 in: Plant Dis. Rep. Suppl. 40.

Starr, M. P., Volcani, Z., and Munnecke, D. E. 1955. Relationship of *Xanthomonas pelargonii* and *Xanthomonas geranii*. Phytopathology 45:613-617.

Bacterial Leaf Spot of *Begonia*

Bacterial leaf spot is an important foliar disease of *Begonia*. Despite a general recognition of the problem, it continues to be common and widespread. Mild symptoms can easily be overlooked, which may explain the regular occurrence of the disease. The foliar nematode, *Aphelenchoides*, increases the severity of the disease when the nematode and the bacterium occur together.

Symptoms

Symptoms vary depending on the species or interspecific cross. On wax begonia (*Begonia* Semperflorens-Cultorum hybrids), lesions about 1 mm in diameter are first apparent on the undersurface of the leaf. As lesions enlarge to 2–8 mm in diameter, they are roughly circular with irregular borders and translucent halos. Symptoms are similar on nonstop begonias, *Begonia* Tuberhybrida hybrids (Plate 163). Affected leaves may fall off, and the bacterium may or may not become systemic and cause wilt. On hiemalis (Rieger) begonias (*B.* × *hiemalis* Fotsch.), lesions often originate on the leaf margins and become brown wedges several centimeters across (Plate 164). The lesions have a characteristic stippling not seen on the wax begonia. Cultivars of hiemalis begonia vary in their susceptibility.

Causal Organism

Bacterial leaf spot of *Begonia* is caused by *Xanthomonas campestris* pv. *begoniae* (Takimoto) Dye, which has the characteristics listed above for *X. c. pelargonii*, except that it is bright yellow on YDC (yeast extract-dextrose-$CaCO_3$) medium and grows well on SX medium.

Host Range and Epidemiology

X. c. begoniae is restricted to the genus *Begonia*. Various species, cultivars, and hybrids of *Begonia* are susceptible. *B. alba*, *B.* × *hiemalis* Fotsch, *B. gracilis* HBK, *B. socotrana* Hook. f., and *Begonia* Rex-Cultorum, Semperflorens-Cultorum, and Tuberhybrida hybrids are reported to be hosts. Hybrids of bulbous-rooted and tuberous types, *B.* × *hiemalis*, are susceptible to various degrees.

Infected plant materials are probably the most important source of contamination. *X. c. begoniae* can survive in dried begonia leaves for at least 1 year. It may also reside on foliage for several months before initiating disease. Survival in water is reportedly short, probably less than several weeks. While it is possible that the bacterium is seedborne, this has not been reported.

Xanthomonas can become systemic in begonia, especially in the wax and hiemalis types. It can move from roots to foliage or from foliage to roots. The bacterium has been shown to be released through roots of diseased plants into ebb-and-flood irrigation systems, resulting in transmission to healthy plants. However, the incidence of such transmission appears to be very low. It is more likely to be spread by conventional irrigation practices that result in splashing water from plant to plant.

Management

Sanitation is an important disease-management principle that is particularly pertinent to bacterial problems such as Xanthomonas leaf spot of begonia. Plants with leaf spots should be discarded because they are likely to be systemically infected. Workers should wash their hands after handling diseased plants or soil. Diseased plant debris should be removed from the growing area as much as possible. Handling of wet foliage should be avoided. Since bacteria are easily splashed from plant to plant by irrigation water, practices, such as ebb-and-flood irrigation, that minimize splashing and reduce leaf wetness duration should be considered. Ebb-and-flood systems should utilize a relatively short flooding duration so that the capillarity of the growing medium is still pulling irrigation water upward when the water is draining. Screens should be installed in the drainage end of the ebb-and-flood bench to prevent infected plant material from recirculating.

Copper hydroxide is registered in the United States for control of *Xanthomonas* on begonia. Although sprays with this bactericide have been shown to reduce the rate of foliar infection, once the bacterium becomes systemic, sprays will be of no help. Affected plants should be discarded.

Selected References

Atmatjidou, V. P., Fynn, R. P., and Hoitink, H. A. J. 1991. Dissemination and transmission of *Xanthomonas campestris* pv. *begoniae* in an ebb and flow irrigation system. Plant Dis. 75:1261-1265.

Jodon, M. H., and Nichols, L. P. 1974. Bacterial leaf spot of begonia. Penn. Flower Growers Bull. 272:8-9.

McCulloch, L. 1937. Bacterial leaf spot of begonia. J. Agric. Res. 54:583-590.

Nichols, L. P., Jodon, M. H., and Scarborough, B. 1974. Longevity of *Xanthomonas begoniae*, the cause of bacterial leafspot of Rieger begonias. Plant Dis. Rep. 58:814.

Riedel, R. M., and Larsen, P. O. 1974. Interrelationships of *Aphelenchoides fragariae* and *Xanthomonas begoniae* on Rieger begonia. J. Nematol. 4:215-216.

Strider, D. L. 1975. Chemical control of bacterial blight of Rieger elatior begonias caused by *Xanthomonas begoniae*. Plant Dis. Rep. 59.66-70.

Strider, D. L. 1975. Susceptibility of Rieger elatior begonia cultivars to bacterial blight caused by *Xanthomonas begoniae*. Plant Dis. Rep. 59:70 73.

Strider, D. L. 1978. Reaction of recently released Rieger elatior begonia cultivars to powdery mildew and bacterial blight. Plant Dis. Rep. 62:22-23.

Xanthomonas Leaf Spot of Poinsettia

Leaf spot of *Euphorbia pulcherrima* Willd. ex Klotzsch (poinsettia) caused by *Xanthomonas campestris* pv. *poinsettiicola* was first described in India in 1951. The disease was subsequently described in Florida in 1962, and in 1985 the pathogen was also reported on another euphorbiaceous plant, *Codiaeum*. Two other pathovars of *X. campestris* have also been reported to cause leaf spot on poinsettia.

Symptoms

Spots are at first visible on the underside of the leaf as gray to brown, water-soaked lesions. As they enlarge to 2–3 mm, they become visible on the top side of the leaf and are chocolate brown to rust colored and may or may not have pale green halos (Plate 165). Spots may coalesce to form large areas of blighted tissue. Severely spotted leaves turn yellow and drop from the plant.

Causal Organisms

X. c. poinsettiicola (Patel, Bhatt, & Kulkarni) Dye has the characteristics described for *X. campestris*. However, this pathovar is known to infect only a few species in the family Euphorbiaceae. *X. c. euphorbiae* (Sabet, Ishag, & Khalil) Dye and *X. c. manihotis* (Berthet & Bondar) Dye have been reported to cause poinsettia leaf spot upon inoculation.

Host Range and Epidemiology

X. c. poinsettiicola is pathogenic to *E. pulcherrima* and *Codiaeum variegatum* (L.) Blume. Both hosts are in the Euphorbiaceae. *X. c. euphorbiae* causes leaf spot on *E. acalyphoides*, a weed from Sudan, and will spot poinsettia upon inoculation. *X. c. manihotis*, a nonpigmented xanthomonad, is pathogenic to *Manihot esculenta* Crantz (cassava) and to poinsettia upon inoculation.

Little has been published on the epidemiology of this disease, but it has been reported to spread rapidly, presumably from splashing water. If epiphytic populations occur, as is known with other diseases caused by xanthomonads, then colonized, nonsymptomatic cuttings may develop disease under favorable conditions.

Management

Affected plants should be discarded, and no cuttings should be taken from plants with leaf spots. Wetting of foliage should be avoided, especially during periods that would extend the leaf wetness duration.

Selected Reference

McFadden, L. A., and Morey, H. R. 1962. Bacterial leaf spot disease of poinsettia in Florida. Plant Dis. Rep. 46:551-554.

Bacterial Spot of *Capsicum*

Bacterial spot of ornamental pepper, *Capsicum annuum* L. var. *annuum*, is a well-known and an important problem of the closely related edible peppers (also *C. annuum*). Bacterial spot is also an important disease of *Lycopersicon esculentum* Mill. (tomato). No research has been published specifically on bacterial spot of ornamental pepper, but this disease has the potential to cause significant economic losses.

Symptoms

Spots on ornamental pepper occur on the foliage and are similar to those that occur on edible peppers. Spots are at first small, circular, and dark green or water soaked in appearance. Individual spots may develop tan centers with water-soaked margins. Large blotches of infected tissue may occur when smaller lesions coalesce. Lesions may also occur on the margins of the leaves (Plate 166). Affected leaves often fall from the plant.

Causal Organism

Xanthomonas campestris pv. *vesicatoria* (Doidge) Dye has the characteristics described for *X. campestris*. This pathovar, however, is restricted to members of the Solanaceae. Several races of *X. c. vesicatoria* have been differentiated on the basis of their ability to infect either pepper or tomato or both.

Host Range and Epidemiology

Hosts of *X. c. vesicatoria* include common vegetables and weeds such as *Capsicum* spp., *Datura stramonium* L., *Hyoscyamus* spp. (henbane), *Lycium* spp. (matrimony-vine), *Nicotiana rustica* L. (Aztec tobacco), *Nicandra physalodes* (L.) Gaertner (apple-of-Peru), *Physalis minima* L. (groundcherry), and *Solanum* spp.

No epidemiological studies have been reported for bacterial spot of ornamental pepper, but the disease has been well characterized on susceptible vegetable crops. *X. c. vesicatoria* is known to be seedborne in both pepper and tomato, and circumstantial evidence strongly indicates that it is seedborne on ornamental peppers as well. Symptoms develop most rapidly when temperatures are between 20 and 30°C. High relative humidity and extended periods of leaf wetness favor disease development. The bacterium can be spread by splashing irrigation water and by handling plants.

Management

Seeds should be purchased from a reputable source, and plants that develop symptoms should be promptly discarded. Splashing water should be avoided. Relative humidity should be reduced to less than 90%, and extended periods of leaf wetness should be avoided.

Selected Reference

Jones, J. B., Jones, J. P., Stall, R. E., and Zitter, T. A., eds. 1991. Compendium of Tomato Diseases. American Phytopathological Society, St. Paul, MN.

Bacterial Leaf Spot
of *Gardenia*

Though reported infrequently, *Xanthomonas campestris* pv. *maculifoliigardeniae* is known to cause a leaf spot on *Gardenia augusta* (L.) Merrill. During the 1940s, heavy losses of gardenia in California were attributed to this leaf spot.

Symptoms

Spots occur on the youngest leaves as small dots. As lesions enlarge, they develop pale yellow centers, which later become brown or reddish brown with yellow halos. The margins of the lesions are thickened and appear greasy. Spots may coalesce to form necrotic blotches. Defoliation often occurs.

Causal Organism

X. c. maculifoliigardeniae (Ark & Barrett) Dye has the characteristics described for *X. campestris*. The known host range of this pathovar is very narrow.

Host Range and Epidemiology

X. c. maculifoliigardeniae has been reported on gardenia and flame-of-the-woods (*Ixora coccinea* L.). No other hosts are known.

Optimum conditions for disease development have not been established, but leaf spots occur within 7–9 days after inoculation at day temperatures of 27–32°C. During the 1940s, the disease was reported to be especially prevalent because of the practice of syringing plants to reduce spider mites, a practice no longer common. The bacterium is spread by vegetative propagation from infected plants and by splashing water.

Management

Affected plants should be discarded, and cuttings should not be taken from plants with leaf spots. Wetting the foliage should be avoided, especially during periods that would extend the duration of leaf wetness.

Selected Reference

Ark, P. A., and Barrett, J. T. 1946. A new bacterial leaf spot of greenhouse-grown gardenias. Phytopathology 36:865-868.

Diseases Caused
by Phytoplasmas

The "yellows" diseases of plants were thought to be caused by viruses and were referred to as aster yellows until 1967,

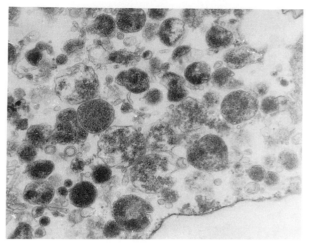

Fig. 48. Phytoplasmas in the phloem of an infected hydrangea. (Courtesy R. Lawson)

when it was discovered that bacteria were the causes. Since the bacteria resembled animal pathogens called mycoplasmas, the new plant pathogens were called mycoplasmalike organisms (MLOs) until 1994. They are now known as phytoplasmas, and the various strains will be classified as members of a new candidate genus, *Phytoplasma*.

Phytoplasmas are linked to diseases of more than 200 species of plants. Flowering potted plants reported to be hosts include *Begonia* spp., *Catharanthus roseus* (vinca), *Cyclamen persicum*, *Dahlia* hybrids, *Gerbera jamesonii* (African daisy), *Hydrangea macrophylla*, *Mimulus* sp. (monkey-flower), *Pericallis* × *hybrida* (cineraria), *Primula* spp., *Ranunculus* hybrids, *Schizanthus* sp. (butterfly flower), and *Sinningia speciosa* (gloxinia).

Typical symptoms include yellowing or bronzing of the foliage, virescence, stunting, sterile flowers, abnormal fruit and seeds, proliferation of roots, and the development of witches'-brooms. Phytoplasmas are vectored by leafhoppers, planthoppers, or psyllids and are restricted to the phloem of infected plants. They can be detected by electron microscopy, but their pleomorphic shape makes them hard to distinguish from other membrane-bound components of plant cells (Fig. 48). Serological methods and DNA-based techniques have received recent emphasis for phytoplasma detection.

Selected Reference

Anonymous. 1960. Index of plant diseases in the United States. U.S. Dep. Agric. Agric. Handb. 165.

Hydrangea Virescence

Hydrangea virescence is a graft-transmissible disease that has been reported from Europe and North America. In the past, virescence symptoms were attributed to hydrangea ringspot virus; more recently, they have been associated with the presence of a phytoplasma. Virescence of *Hydrangea macrophylla* (Thunb.) Ser. is also known as phyllody or proliferation.

Symptoms

The name of the disease comes from the predominant symptom: some of the cymes on infected hydrangeas have normal color, while others have all green (virescent) florets (Plate 167). In the most severe cases, the cymes are virescent; the florets are stunted; and leafy, shootlike structures grow out from the pistils, while leaves are stunted with yellow veins. These symptoms increase in severity over time, and the plants

58

eventually succumb. Plants with milder strains of the pathogen will not die but will continue to show symptoms annually.

Causal Organism

Phytoplasmas are pleomorphic, membrane-bounded bacteria without rigid cell walls. They cannot be cultured in the laboratory; therefore, little is known about their physiology. More than one phytoplasma may be responsible for the complex of symptoms known as hydrangea virescence. Some strains detected in diseased hydrangea have been identified as members of a large genomic cluster containing strains of aster yellows phytoplasma.

Host Range and Epidemiology

Host specificity of phytoplasmas is not completely understood. Attempts to transfer hydrangea virescence to *Callistephus chinensis* (L.) Nees (aster) or *H. macrophylla* via the aster leafhopper (*Macrosteles fascifrons* Stål) have failed. Three forms of the hydrangea virescence agent (associated with three disease severity levels) have been noted, each of which consistently produces the same pattern of symptoms in graft-inoculation trials. The mildest form, marked only by flower symptoms, is the hardest to eliminate by roguing, since symptom expression is delayed until the crop is in the finishing stage. The phytoplasma is graft transmissible in hydrangea but not mechanically transmitted. No vectors have been identified for hydrangea virescence.

Management

New hybrids should be grown separately from older cultivars, since isolation is the best defense when the vectors of a disease are unknown. The disease is not mechanically transmitted, which reduces the likelihood of its spread during cultural practices. Plant spacing could be important for disease control in field-grown stock because root grafting might allow transmission between plants. Nuclear stock of hydrangeas free from viruses and the virescence agent can be developed by meristem tip culture or by heat therapy followed by meristem culture.

Selected References

Davis, R. E., and Lee, I.-M. 1982. Pathogenicity of spiroplasmas, mycoplasmalike organisms, and vascular-limited fastidious walled bacteria. Pages 491-513 in: Phytopathogenic Prokaryotes, vol. 1. M. Mount and G. Lacy, eds. Academic Press, New York.

Hearon, S. S., Lawson, R. H., Smith, F. F., McKenzie, J. T., and Rosen, J. 1976. Morphology of filamentous forms of a mycoplasmalike organism associated with hydrangea virescence. Phytopathology 66:608-616.

Lawson, R. H. 1985. Hydrangea. Pages 259-273 in: Diseases of Floral Crops, vol. 2. D. L. Strider, ed. Praeger Scientific, New York.

Lawson, R. H., and Smith, F. F. 1980. Three stable types of hydrangea virescence that differ in severity in foliage and flowers. Plant Dis. 64:659-661.

Diseases Caused by Viruses

Plant viruses are simple organisms consisting of a few genes composed of nucleic acid packaged in a protein capsule (capsid). They are very small (20–300 nm) and visible only by electron microscopy (Fig. 49). Aggregates of virus particles, called inclusion bodies, can sometimes be seen within plant cells with a light microscope (Plate 168). Viruses disrupt normal plant functions by utilizing the plant's structural and chemical resources to manufacture more virus particles. The result of this redirection of the plant's metabolism is plant disease. Viruses do not usually kill plants, but they often reduce their ornamental value. The name of a virus describes one of the symptoms it may cause and gives the name of one of the plants that it is able to infect (usually the first host on which it was described).

Viruses cause a diversity of symptoms, some of which are unique to viruses but many of which mimic those of nutrient disorders, herbicide injury, insect damage, or other plant pathogens such as fungi or bacteria. Some of the plant diseases that once were attributed to viruses are now known to be caused by other kinds of organisms, such as phytoplasmas (previously termed mycoplasmalike organisms [MLOs]), viroids, and xylem limited bacteria.

Symptom expression is the result of a virus's redirection of normal plant metabolism. The most common symptom is stunting, which often goes unnoticed when virus-free plants are not available for comparison. Other virus infection symptoms include color patterns such as mosaics or streaks; leaf chlorosis; veinclearing; veinbanding; ring spots; necrotic spots, streaks, stem tips, or veins; distorted leaves; swollen shoots; enations; and flower break. Viruses can also affect reproduction and cause reduced flower number, fruit reduction or distortion, reduced seed set, pollen sterility, or reduced viability. In some instances, plants may wilt and die. Diagnostic species (indicator plants) used for virus identification exhibit distinctive local lesions and/or systemic symptoms when inoculated with different viruses (Plate 169 and Fig. 50).

The development of virus-free plants is an important concern, particularly for vegetatively propagated crops such as chrysanthemum, geranium, and carnation. In order to develop a virus-free "mother block" of propagation stock, heat therapy prior to meristem tip culture and virus-indexing are used. Indexing is carried out on indicator plants or by serological techniques such as ELISA (enzyme-linked immunosorbent assay).

Seventeen viruses of flowering potted plant crops are described in detail in the following text. Some of these are known to occur only on ornamentals, while others affect a wide range of horticultural crops and weed species. Additional viruses reported from flowering potted plants, but which are generally not of great economic importance on these crops, are listed in Table 17.

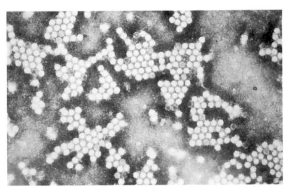

Fig. 49. Particles of poinsettia mosaic virus viewed by transmission electron microscopy. (Courtesy S. Nameth)

Fig. 50. Inoculating indicator plants with infected plant sap allows characteristic plant responses that aid in the identification of viruses. (Courtesy M. Klopmeyer)

Selected References

Matthews, R. E. F. 1991. Plant Virology. 3rd ed. Academic Press, San Diego.

Walkey, D. G. A. 1991. Applied Plant Virology. 2nd ed. St. Edmundsbury Press, Edmunds, England.

Alfalfa Mosaic of Geranium, *Hydrangea,* and *Primula*

Alfalfa mosaic virus (AMV) has been reported infrequently in greenhouse flower crops. Its wide host range and common vectors, however, suggest that it may be more widespread than reports would indicate.

Symptoms

Primula sp. (primrose) infected with AMV are stunted. Young leaves exhibit a mosaic pattern, while older leaves have yellow patches and necrotic spots. Color breaking is evident in the flowers. During the summer, these symptoms recede. *Pelargonium* spp. (geraniums) infected with AMV may show yellow interveinal patches. The virus has also been detected in *Hydrangea macrophylla* (Thunb.) Ser. AMV is symptomless in many of its hosts.

Causal Agent

AMV has a tripartite genome divided into three bacilliform particles (36, 48, and 58 × 18 nm). A separate spherical particle (18 nm) carries the coat protein messenger RNA. Plant infection requires all four particles or the three largest particles plus some coat protein.

Host Range and Epidemiology

AMV has been reported on 150 plant species in 22 families and is distributed worldwide. It is transmitted in a nonper-sistent fashion by aphids, including *Myzus persicae* (Sulzer), the green peach aphid, and has been transmitted experimentally to plants in more than 51 different families. Seed transmission is known for some hosts but has not been demonstrated for any of the flowering potted plants.

Diagnostic species—*Chenopodium amaranticolor* Coste & Reyn. (pigweed) and *C. quinoa* Willd. (quinoa): systemic chlorotic and necrotic flecks separate alfalfa mosaic from cucumber mosaic; *Nicotiana tabacum* L. (tobacco): chlorotic or necrotic local lesions; *Ocimum basilicum* L. (sweet basil): systemic yellow mosaic; *Phaseolus vulgaris* L. (common bean): local lesions or mottle, vein necrosis, and leaf distortion, depending on virus strain; *Pisum sativum* L. (pea): local lesions, wilting of inoculated leaves, stem necrosis, and death; *Vicia faba* L. (broad bean): usually black, necrotic local lesions followed by a mild mottle or stem necrosis and death; *Vigna unguiculata* (L.) Walp. (cowpea): necrotic local lesions from common virus strains. In tobacco, transient, amorphous, granular inclusion bodies occur.

Management

Control of AMV is achieved by weed management and control of its aphid vectors during both seed production and finished crop production. In cases in which propagation is done vegetatively (such as with *Hydrangea* and *Pelargonium*), clean stock programs should incorporate screening for AMV.

Selected References

Betti, L., and Canova, A. 1981. Pelargonium yellow flaking caused by Alfalfa mosaic virus. Phytopathol. Mediterr. 20:176-178.

Jaspers, E. M. J., and Bos, L. 1980. Alfalfa mosaic virus. Descriptions of Plant Viruses, no. 229. Commonwealth Mycological Institute and Association of Applied Biologists, Kew, England.

Nagaich, B. B., and Giri, B. K. 1968. A strain of alfalfa mosaic virus from primula and potato. Indian Phytopathol. 23:446-447.

Bean Yellow Mosaic of Lisianthus

Symptoms

Eustoma grandiflorum (Raf.) Shinn. (lisianthus) infected with bean yellow mosaic potyvirus (BYMV) shows mosaic, leaf curl, chlorotic spots, and flower break. If tobacco mosaic or cucumber mosaic is present as well, the plants are also stunted. When older plants are inoculated, milder symptoms are produced.

Causal Agent

BYMV consists of flexuous rods 700–760 nm in length that contain single-stranded RNA. The potyviruses are transmitted by aphids in a nonpersistent manner and by sap inoculation. Individual viruses in this genus have relatively restricted host ranges. Some of the members are seed transmitted. Inclusion bodies observable within the epidermis are helpful for diagnosis. Cytoplasmic inclusions are granular or crystalline, and there may also be crystalline inclusions in the nucleolus.

Diagnostic species—*Phaseolus vulgaris* L. (common bean): inoculated primary leaves often show epinasty and develop local lesions, systemic yellow mosaic, leaf distortion, and stunting; *Pisum sativum* L. (pea): transient vein chlorosis followed by mild to yellow mosaic; *Vicia faba* L. (broad bean): transient vein chlorosis followed by obvious mosaic; *Chenopodium amaranticolor* Coste & Reyn. (pigweed): chlorotic local lesions, systemic vein yellowing, and leaf malformation; *C. quinoa* Willd. (quinoa): chlorotic local lesions; *Gomphrena globosa* L. (globe-amaranth): necrotic local lesions; *Nicotiana tabacum* L. (tobacco), *Petunia* × *hybrida* Vilm., and *Tetragonia expansa* Thunb. ex J. A. Murray (New Zealand spinach): chlorotic local lesions occasionally; *Spinacia oleracea* L. (spinach): local lesions followed by systemic infection.

Host Range and Epidemiology

In addition to lisianthus, BYMV is known to affect several other ornamentals, notably *Gladiolus* × *hortulanus* L. H. Bailey (gladiolus) and *Freesia* × *hybrida* L. H. Bailey. More than 20 aphid vectors are known to transmit this virus in a nonpersistent manner, including *Aphis gossypii* Glover, the melon aphid.

Management

Aphid vector exclusion and virus-free seed are important for keeping BYMV out of the greenhouse. Aphid control will reduce virus spread within the greenhouse.

Selected References

Gera, A., and Cohen, J. 1990. The natural occurrence of bean yellow mosaic, cucumber mosaic and tobacco mosaic viruses in lisianthus in Israel. Plant Pathol. 39:561-564.

Lisa, V., and Dellavalle, G. 1987. Bean yellow mosaic virus in *Lisianthus russelianus*. Plant Pathol. 36:214-215.

TABLE 17. Viruses of Minor Importance on Flowering Potted Plants

Host	Virus	Symptoms
Anemone coronaria	Anemone mosaic	Flower break; distortion; leaf mottling
Begonia spp.	Carnation mottle carmovirus (CarMV)	Veinclearing; leaf curl; deformed flowers; color break
	Tobacco necrosis necrovirus (TNV)	Stunt; poor vigor; ring spots; color variations
	Clover yellow mosaic potexvirus (ClYMV)	Bright mosaic; stunt
Campanula carpatica (bellflower)	Unnamed potyvirus	Symptomless
Catharanthus roseus (vinca)	Periwinkle chlorotic stunt virus	Leaf mosaic and crinkle; stunt
	Periwinkle mosaic	Mosaic
Cyclamen persicum	Tomato aspermy cucumovirus (TAV)	Flower malformation
	Potato X potexvirus (PVX) + tobacco mosaic tobamovirus (TMV)	Flower distortion; necrotic streaking of petals
Dahlia hybrids	Potato Y potyvirus (PVY)	Not known
Euphorbia pulcherrima (poinsettia)	Crinkle mosaic of poinsettia	Thin stems; patchy, bright yellow mosaic; distortion
	Poinsettia cryptic virus	Symptomless
	Enation leaf curl of poinsettia	Leaf curl; stunted, twisted petioles; thick leaves; swollen veins; enations; failure to flower
Eustoma grandiflorum (lisianthus)	Lisianthus necrosis necrovirus	Systemic necrotic spots; rings on leaves and stem; stunt; tip necrosis; flower color break
Gerbera jamesonii (African daisy)	Tobacco rattle tobravirus (TRV)	Black or yellow ring spots
Hibiscus rosa-sinensis	Hibiscus latent ringspot nepovirus (HLRSV)	Symptomless or faint leaf chlorosis
	Unnamed rhabdovirus	Vein yellowing
Hydrangea macrophylla	Hydrangea latent carlavirus	Symptomless
	Hydrangea mosaic ilarvirus (HdMV)	Chlorotic foliar mosaic
	Tobacco necrosis necrovirus (TNV)	Stunt; ring spots; color variations; poor vigor
Impatiens wallerana	Helenium S carlavirus (HVS)	Symptomless or stunt
	Unnamed potexvirus	Symptomless
Kalanchoe blossfeldiana	KV-1 potexvirus and KV-2 carlavirus	Symptomless
	Unnamed potyvirus	Symptomless
Pelargonium spp. (geranium)	Arabis mosaic nepovirus (ArMV)	Mild mosaic; chlorosis
	Artichoke Italian latent nepovirus (AILV)	Not known
	Beet curly top geminivirus (BCTV) (= Pelargonium leaf curl)	Inwardly cupped leaves; loss of pubescence
	Moroccan pepper tombusvirus (MPV)	With pelargonium flower-break carmovirus: leaf malformation; stunt
	Pelargonium ring pattern virus (PRPV) (possibly = Pelargonium line pattern)	Yellow spots; rings; necrotic line patterns
	Pelargonium veinclearing virus (PVCV)	Veinclearing
	Pelargonium zonate spot virus (PZSV)	Concentric, chrome yellow bands
	Tobacco necrosis necrovirus (TNV)	Not known
	Tobacco rattle tobravirus (TRV)	Not known
	Tomato black ring nepovirus (TBRV)	Not known
Primula spp. (primrose)	Tobacco necrosis necrovirus (TNV)	Mottle; chlorosis; stunt; necrotic spots along veins
	Tomato bushy stunt tombusvirus (TBSV)	Systemic necrosis
	Primula mosaic	Mosaic; green, elevated blisters; upcupping of young leaves; shoestring leaves; petal spots and streaks; stunt
	Primula mottle potyvirus	Mild leaf mosaic; stunt; flower break
Ranunculus asiaticus	Potato Y potyvirus (PVY)	Symptomless; but in mixed infections plants show marked leaf chlorosis and deformation
	Ranunculus potyvirus (RV)	Severe systemic foliar mottling; stunt
Verbena × *hybrida*	Bidens mottle potyvirus (BiMoV)	Stunt; mottling; leaf distortion

Broad Bean Wilt
of *Begonia* and Lisianthus

Broad bean wilt virus (BBWV) has been seen in only a few flowering potted plants but is potentially very destructive.

Symptoms

BBWV in *Begonia* Semperflorens-Cultorum hybrids (wax begonia) causes symptoms clearly different from those caused by the more common tobacco ringspot virus. Begonias infected with BBWV show leaf mottling, faint ring spots, and stunting, all of which are more pronounced during the winter months. In pink- and red-flowered varieties, color breaking also occurs. This virus has also been reported in *Eustoma grandiflorum* (Raf.) Shinn. (lisianthus).

Causal Agent

BBWV is a member of the genus *Fabavirus,* whose members each possess a bipartite genome of single-stranded RNA packaged in isometric particles, which sediment as three separate components. Fabaviruses are transmitted in a nonpersistent manner by aphids and may be sap transmitted. They have a wide experimental host range. BBWV has isometric particles 25 nm in diameter. Depending on the virus strain, various types of inclusion bodies may be observed with a light microscope.

Host Range and Epidemiology

Several strains of BBWV have been reported from countries throughout the world. The virus has a wide host range and is transmitted by *Myzus persicae* (Sulzer), the green peach aphid.

Diagnostic species—*Vicia faba* L. (broad bean): veinclearing followed by terminal leaf necrosis, wilting, and often death; *Datura stramonium* L. (jimson-weed): local lesions in the form of concentric necrotic rings, ring spots, and line patterns on new leaves; *Chenopodium quinoa* Willd. (quinoa): chlorotic local lesions followed by chlorosis of the whole plant or the apical leaves; *Nicotiana tabacum* L. cv. White Burley (tobacco): chlorotic ring spots that enlarge and become necrotic.

Management

Control of the green peach aphid will limit the spread of the virus and reduce the chance that the disease will be introduced to begonia or other susceptible crops from weed crops in or near the greenhouse.

Selected References

Iwaki, M., Maria, E. R. A., Hanada, K., Onogi, S., and Zenbayashi, R. 1985. Three viruses occurred in lisianthus plants. Ann. Phytopathol. Soc. Jpn. 52:355.

Lockhart, B. E. L., and Betzold, J. A. 1982. Broad bean wilt virus in begonia in Minnesota. Plant Dis. 66:72-73.

Taylor, R. H., and Stubbs, L. L. 1972. Broad bean wilt virus. Descriptions of Plant Viruses, no. 81. Commonwealth Mycological Institute and Association of Applied Biologists, Kew, England.

Cucumber Mosaic

Cucumber mosaic virus (CMV) is the type member for the genus *Cucumovirus.* CMV's wide host range, worldwide distribution, and number of aphid vectors make it a very commonly encountered virus. It may produce very striking symptoms in some hosts.

Symptoms

A number of flower crops show dramatic symptoms when infected with CMV—*Begonia:* stunting and ring spots; *Campanula, Catharanthus,* and *Cyclamen persicum* Mill.: flower streaking and distortion; *Dahlia:* interveinal mosaic; *Eustoma grandiflorum* (Raf.) Shinn. (lisianthus): foliar mosaic, leaf necrosis, leaf distortion, severe stunting, flower break, and malformation; *Gerbera jamesonii* H. Bolus ex Adlam (African daisy): petal break, flower distortion, stunting, and mottling of leaves; *Hydrangea macrophylla* (Thunb.) Ser. (hydrangea) and *Impatiens* sp.: leaf curl; New Guinea impatiens: leaf strapping and leaf curl (Plate 170); *Pelargonium* spp. (geranium): leaf breaking and mosaic; *Primula obconica* Hance (German primrose) and *P. sinensis* Sabine ex Lindl. (Chinese primrose): necrosis, stunting, mottling of leaves, veinclearing, mosaic, green islands imbedded in yellow leaf tissue, and altered flower color; and *Solanum pseudocapsicum* L. (Jerusalem cherry): leaf deformation, ringlike mottling, and stunting.

In some cases, CMV alone causes symptoms in plants. It may also contribute to the expression of symptoms in mixed infections with other viruses. There are many strains of CMV, and they vary in the symptoms they produce.

Causal Agent

Cucumoviruses (including CMV) have tripartite genomes with three RNA-containing particles, each approximately 28 nm, which are all necessary for infection. The particular virus strain or the presence of satellite RNA affects the host response.

Host Range and Epidemiology

CMV may be experimentally sap transmitted to at least 191 species in 40 families. It is transmitted in a nonpersistent manner by more than 60 species of aphids, including green peach (*Myzus persicae* (Sulzer)) and melon (*Aphis gossypii* Glover) aphids. Aphids can acquire the virus within 5–10 sec. Vectors typically transmit the virus most efficiently for the first 2 min after acquisition and generally for no longer than 2 hr. Aphid transmission of CMV is unlikely for geraniums, since they are not attractive to the vector. Seed transmission is known for 19 host species.

Diagnostic species—*Chenopodium amaranticolor* Coste & Reyn. (pigweed) and *C. quinoa* Willd. (quinoa): chlorotic or necrotic local lesions; *Cucumis sativus* L. (cucumber) and *Cucurbita moschata* (Duchesne) Duchesne ex Poir. (pumpkin): systemic mosaic and stunt; *Lycopersicon esculentum* Mill. (tomato): mosaic and narrowed leaf blades; *Nicotiana glutinosa* L. (tobacco): mild to severe mosaic; and *Vigna unguiculata* (L.) Walp. (cowpea): small brown local lesions. Inclusion bodies are easily detected in some hosts but have been reported to be ephemeral in others.

Management

Symptomatic plants should be rogued. Weeds in and around the greenhouse should be eliminated, and aphid populations should be controlled. Virus-indexing for CMV should be carried out when virus-free stock of any ornamental that is a potential host is created.

Selected References

Francki, R. I. B., Mossop, D. W., and Hatta, T. 1979. Cucumber mosaic virus. Descriptions of Plant Viruses, no. 213. Commonwealth Mycological Institute and Association of Applied Biologists, Kew, England.

Gera, A., and Cohen, J. 1990. The natural occurrence of bean yellow mosaic, cucumber mosaic and tobacco mosaic viruses in lisianthus in Israel. Plant Pathol. 39:561-564.

Dahlia Mosaic

Symptoms

On many *Dahlia* hybrid cultivars infected with dahlia mosaic caulimovirus (DMV), chlorotic bands develop along the midribs and larger veins of the leaves followed by mosaic and leaf distortion. The most sensitive cultivars are stunted and have a bushy habit. Tubers are also shortened and thickened. Some cultivars, however, are nearly symptomless.

Causal Agent

DMV is a member of the genus *Caulimovirus*. The caulimoviruses are unique in that they have genomes of double-stranded DNA (most plant viruses have an RNA genome). Caulimoviruses have isometric particles approximately 50 nm in diameter. The host range is limited. Caulimoviruses are transmitted by aphids in a nonpersistent manner, and some are not easily sap transmitted. DMV has spherical, DNA-containing particles that are 50 nm in diameter. Spherical or ellipsoidal inclusion bodies may be seen by light microscopy, especially in epidermal, palisade, and spongy parenchyma cells.

Host Range and Epidemiology

DMV has a worldwide distribution. Only *Dahlia* is naturally infected, but the virus has been experimentally sap transmitted to other composites and to certain members of the Solanaceae, Chenopodiaceae, and Amaranthaceae.

Diagnostic species—*Verbesina encelioides* (Cav.) A. Gray (golden-crown daisy): chlorotic local lesions followed by chlorotic veinbanding and general chlorosis; *Dahlia pinnata* Cav.: systemic chlorotic veinbanding 3 weeks or longer after inoculation, chlorosis, and stunting; *Ageratum conyzoides* L. cv. Blue Ball: chlorotic local lesions followed by systemic veinbanding and chlorosis; *Zinnia elegans* Jacq.: systemically infected leaves slightly distorted, transient chlorosis, and stunting; *Amaranthus caudatus* L. (love-lies-bleeding): sparse chlorotic local lesions followed by light systemic mottling; and *Chenopodium capitatum* (L.) Asch. (Indian-paint): no local lesions but systemic chlorotic veinbanding and chlorosis.

Sap transmission to dahlia is difficult. The virus is vectored in a nonpersistent manner by the green peach aphid, *Myzus persicae* (Sulzer). Twelve other aphid species have also been shown to transmit DMV.

Management

Infected dahlia stock should be discarded. Aphid exclusion and control are important for prevention of disease spread from symptomless cultivars or possible undiscovered weed hosts.

Selected Reference

Brunt, A. A. 1971. Dahlia mosaic virus. Descriptions of Plant Viruses, no. 51. Commonwealth Mycological Institute and Association of Applied Biologists, Kew, England.

Hibiscus Chlorotic Ringspot

Symptoms

Hibiscus chlorotic ringspot carmovirus (HCRSV) occurs in many cultivars of *Hibiscus rosa-sinensis* L. around the world. Symptoms include systemic mosaics, tiny chlorotic spots, and occasional rings. No flower symptoms are known. Symptoms may appear during the spring and disappear during summer months.

Causal Agent

HCRSV belongs to the genus *Carmovirus*, whose members have a monopartite, single-stranded RNA genome and two subgenomic RNAs that are encapsidated in 30-nm isometric particles. These viruses tend to have a wide host range. Carmoviruses are sap transmissible. In nature, they are vectored by Coleoptera (nonpersistent transmission) or are trans-

mitted without the aid of a vector. No vector has been reported for HCRSV. HCRSV has single-stranded RNA encased in three spherical particles that are 27–30 nm in diameter. Host-associated variants have been reported for this virus. No crystalline inclusion bodies occur in infected cells.

Host Range and Epidemiology

HCRSV is easily transmitted by pruning shears and by grafting. No seed transmission or insect vectors are known. The virus has a worldwide distribution and is common in Florida. It causes systemic symptoms only in members of the Malvaceae.

Diagnostic species—*Chenopodium quinoa* Willd. (quinoa): chlorotic local lesions; *Cyamopsis tetragonoloba* (L.) Taub. (cluster bean), *Macrotyloma uniflorum* (Lam.) Verde (horse gram), and *Gossypium hirsutum* L. (upland cotton): necrotic local lesions; *Hibiscus cannabinus* L. (kenaf): local lesions and systemic mottle; and *Phaseolus vulgaris* L. cv. Pinto (pinto bean): pinpoint local lesions.

Management

Plants should be obtained from seed or virus-indexed meristem tip culture. Contamination of clean stock by mechanical transmission from infected landscape plants or nonindexed cultivars should be avoided.

Selected References

Hurtt, S. S. 1987. Detection and comparison of electrophorotypes of hibiscus chlorotic ringspot virus. Phytopathology 77:845-850.

Raju, B. C. 1985. Occurrence of chlorotic ringspot virus in commercial *Hibiscus rosa-sinensis*. Acta Hortic. 164:273-280.

Waterworth, H. 1980. Hibiscus chlorotic ringspot virus. Descriptions of Plant Viruses, no. 227. Commonwealth Mycological Institute and Association of Applied Biologists, Kew, England.

Hydrangea Ringspot

Hydrangea ringspot potexvirus (HRSV) has been recognized in ornamental hydrangeas (*Hydrangea macrophylla* (Thunb.) Ser.) since the 1950s. Cultivars tolerant of this virus, showing few or no symptoms, have been favored by the greenhouse industry.

Symptoms

The symptoms produced by HRSV on inoculated hydrangeas are chlorotic to brown leaf spots or rings followed by leaf distortion and buckling several months later (Plate 171). Many hydrangeas carry the virus asymptomatically; others show dramatic crinkling, leaf roll and asymmetry, floret number reduction, and stunting. The symptoms may vary with cultivar and environmental conditions. Symptoms tend to be most apparent during the early spring and late summer.

Causal Agent

HRSV consists of short, wide, flexuous rods (490 nm in length) that probably contain single-stranded RNA. Its physical properties place it in the genus *Potexvirus*. It has a slight serological relationship to some of the other potexviruses. Unlike the other viruses common on hydrangea, it is not capable of infecting solanaceous plants.

Diagnostic species—*Gomphrena globosa* L. (globe-amaranth): local necrotic dots that eventually enlarge and acquire red margins; *Chenopodium quinoa* Willd. (quinoa): local chlorotic spots, sometimes developing into chlorotic ring spots; and *Chenopodium amaranticolor* Coste & Reyn. (pigweed): local chlorotic dots, each with a necrotic speck at the center. No inclusion bodies are found within hydrangea, but chloroplasts may appear clumped.

Host Range and Epidemiology

HRSV is easily sap transmitted when cuttings are taken. No vector is known; experimental attempts at aphid transmission have failed. The disease has been found in the United States, Europe, and New Zealand and is thought to be worldwide in cultivated hydrangea, *H. macrophylla,* which is the only natural host known.

Management

Since this virus may produce severe symptoms on some cultivars of hydrangea, it should be indexed for and eliminated from production stock through meristem tip culture or heat treatment. *G. globosa, C. quinoa,* and *C. amaranticolor* are the best indicator plants. Some cultivars have been successfully freed of the virus by heat treatment.

Selected Reference

Koenig, R. 1973. Hydrangea ringspot virus. Descriptions of Plant Viruses, no. 114. Commonwealth Mycological Institute and Association of Applied Biologists, Kew, England.

Kalanchoe Top Spotting

Top spotting has been observed commonly on *Kalanchoe blossfeldiana* Poelln. in commercial greenhouses in Minnesota. *K. daigremontiana* Hamet & Perrier and *K. fedtschenkoi* Hamet & Perrier are also hosts.

Symptoms

Young leaves of kalanchoe infected with kalanchoe top-spotting badnavirus (KTSV) develop chlorotic spots that later become depressed. The symptoms develop during the spring and fall. With care, symptoms can be distinguished from those of kalanchoe mosaic potyvirus (KMV), described in Denmark (Plate 172), which typically causes a green island mosaic.

Causal Agent

KTSV is in the genus *Badnavirus,* which contains non-enveloped, bacilliform viruses with double-stranded DNA. Virus particles are 20–25 × 50–100 nm. KTSV has been found to cause top spotting in cultivars that also contain KV-1, a symptomless potexvirus, but the presence of KV-1 is not necessary for symptoms of top spotting to develop.

Host Range and Epidemiology

KTSV has a very limited host range. It has been mechanically transmitted to several *Kalanchoe* spp., but transmission attempts were unsuccessful on *Chenopodium quinoa* Willd. (quinoa), *Nicotiana benthamiana* Domin and *N. glutinosa* L. (tobaccos), *Cucumis sativus* L. (cucumber), *Pisum sativum* L. (pea), and *Phaseolus vulgaris* L. (common bean). KTSV is transmitted by mealybugs, particularly *Planococcus citri* (Risso). Transmission via grafting, pollen, and seed has also been demonstrated. KTSV is serologically unrelated to other badnaviruses.

Management

Care should be taken to free plant material of viruses before utilizing it in a breeding program. Only virus-free stock should be used for seed production.

Selected References

Hearon, S. S., and Locke, J. C. 1984. Graft, pollen, and seed transmission of an agent associated with top spotting in *Kalanchoë blossfeldiana.* Plant Dis. 68:347-350.

Husted, K., Bech, K., Albrechtsen, M., and Borkhardt, B. 1994. Identification, partial sequencing, and detection of a potyvirus

from *Kalanchoë blossfeldiana.* Phytopathology 84:161-166.

Lockhart, B. E. L., and Ferji, Z. 1988. Purification and transmission of kalanchoe top-spotting-associated virus. Acta Hortic. 234:73-77.

Pelargonium Flower Break

Pelargonium flower break carmovirus (PFBV) was originally discovered in England in *Pelargonium* × *domesticum* L. H. Bailey, the regal or Martha Washington geranium. It occurs naturally only in *Pelargonium* and presently appears to be one of the most common viruses in the florist's geranium, *P.* × *hortorum* L. H. Bailey, in both the United States and Europe.

Symptoms

Symptoms of PFBV usually appear during the spring. *P.* × *hortorum* may exhibit a line pattern or ring spots (Plates 173 and 174), and *P. peltatum* (L.) L'Hér. (ivy geranium) may show ring spots. When plants of the very PFBV-sensitive cultivar Springtime Irene are infected, they develop 2- to 5-mm, enlarging, chlorotic spots, primarily in the younger leaves, as well as flower break (white streaking and rugosity of petals) (Plate 175). Chlorotic veinclearing and leaf deformities are also observed. These symptoms are enhanced when there is simultaneous infection with tomato ringspot virus. Infection with the two viruses increases the degree of leaf deformity and intensifies the color of chlorotic streaks and ring patterns.

Causal Agent

PFBV is in the genus *Carmovirus.* A description of carmoviruses is included in the section on hibiscus chlorotic ringspot. PFBV has isometric particles 27–34 nm in diameter containing single-stranded RNA. Inclusion bodies have not been described.

Host Range and Epidemiology

PFBV occurs in *P.* × *domesticum, P.* × *hortorum,* and *P. peltatum.* The virus has a very restricted experimental host range. It has been transmitted by sap inoculation to only 15 plant species. The virus can be transmitted from geranium to geranium with a contaminated cutting knife, via surface-contaminated pollen, or by thrips in the presence of infested pollen. PFBV may also be spread through recirculating nutrient solution. It is not transmitted by aphids, seed, or dodder (*Cuscuta* spp.). The virus occurs at high concentrations in some cultivars but may be present at very low concentrations in others. In the Netherlands, detection with enzyme-linked immunosorbent assay (ELISA) was found to be reliable year-round in the sensitive cultivar Springtime Irene. Because the virus is unevenly distributed within plants, samples including a mixture of root, leaf, and flower tissue are desirable for diagnosis by ELISA. Immunosorbent electron microscopy has also been used to detect PFBV in geraniums throughout the year. Symptom development after inoculation may take longer than 1 year.

Diagnostic species—PFBV does not cause symptoms when inoculated to *Nicotiana tabacum* L. (tobacco). *Chenopodium quinoa* Willd. (quinoa): chlorotic and necrotic local lesions, occasional systemic infection, and high virus titer; *Gomphrena globosa* L. (globe-amaranth): sparse, brown, necrotic local lesions; *N. clevelandii* Gray. (tobacco): diffuse chlorotic, sometimes necrotic, local lesions; and *Tetragonia expansa* Thunb. ex J. A. Murray (New Zealand spinach): few chlorotic local lesions.

Management

Geraniums have been freed of PFBV by heat treatment. However, the plants are severely stressed by the treatment, and it may be difficult to obtain viable cuttings at the end of the 4-week, 35°C treatment necessary to accomplish the therapy. PFBV was reported to be insensitive to virus-inactivating treatments employing 2% formaldehyde, trisodium orthophosphate, ultraviolet light, or ultrasound. Once virus-free material is obtained, it may be propagated in a clean stock program. In recirculating nutrient solution, slow sand filtration will slow and reduce, but not eliminate, spread of PFBV. In order to remove viruses from drain water, it may be necessary to employ ultrafiltration or heat treatment (above 90°C). Since PFBV has been shown to be spread among geraniums during plant production, virus-free geraniums should not be grown with PFBV-infected geraniums.

Selected References

Bouwen, I., and Maat, D. Z. 1992. Pelargonium flower-break and pelargonium line pattern viruses in the Netherlands; purification, antiserum preparation, serological identification, and detection in pelargonium by ELISA. Neth. J. Plant Pathol. 98:141-156.

Hollings, M., and Stone, O. M. 1974. Pelargonium flower break virus. Descriptions of Plant Viruses, no. 130. Commonwealth Mycological Institute and Association of Applied Biologists, Kew, England.

Krczal, G., Albouy, J., Damy, I., Kusiak, C., Deogratias, J. M., Moreau, J. P., Berkelmann, B., and Wohanka, W. 1995. Transmission of Pelargonium flower break virus (PFBV) in irrigation systems and by thrips. Plant Dis. 79:163-166.

Paludan, N. 1976. Virus diseases in Pelargonium hortorum, especially concerning tomato ringspot virus. Acta Hortic. 59:119-130.

Paludan, N., and Begtrup, J. 1987. Pelargonium × hortorum. Pelargonium flower break virus and tomato ringspot virus: Infection trials, symptomatology and diagnosis. Dan. J. Plant Soil Sci. 91:183-194.

Stone, O. M., and Hollings, M. 1973. Some properties of pelargonium flower-break virus. Ann. Appl. Biol. 75:15-23.

Poinsettia Mosaic

Poinsettia mosaic virus (PnMV) has been reported from greenhouse-grown poinsettias (*Euphorbia pulcherrima* Willd. ex Klotzsch) in the United States, Germany, England, and British Columbia. It appears to be widely distributed in cultivated poinsettias in North America.

Symptoms

PnMV causes mottling or an angular leaf mosaic on poinsettia, but symptoms are expressed only under cool growing conditions (Plate 176). Poinsettias often carry poinsettia cryptic virus as well as PnMV. Poinsettia cryptic virus does not produce any symptoms on poinsettia, and it occurs in low concentration. A chlorotic spotting of unknown cause has also been observed on poinsettias known to be free of PnMV and poinsettia cryptic virus. Leaf and bract distortion is also frequently observed on poinsettia, but this viruslike symptom has not been shown to be caused by PnMV. Since distortion occurs on PnMV-infected poinsettias but has not been seen on plants known to be free of PnMV, it may be that this virus interacts with some other biotic or abiotic agent to create distortion.

Causal Agent

PnMV is possibly a tymovirus. Tymoviruses have a monopartite, single-stranded RNA genome within isometric particles 28–30 nm in diameter. The host range of this group is primarily limited to cruciferous plants; those affecting noncrucifers have narrow host ranges. These viruses are sap transmissible and in nature are transmitted by flea beetles. PnMV has two particles, 26 and 29 nm in diameter, the smaller of which is devoid of RNA (Fig. 49). PnMV resembles the members of the tymovirus group, except for slight morphological differences and a variation in the way it alters the host cell chloroplasts, but has no serological relationship to any of the known tymoviruses. It occurs at a high concentration in host tissue.

Diagnostic species—*Nicotiana benthamiana* Domin (tobacco): inoculated leaves are symptomless, but systemically infected leaves show irregular chlorosis and crinkling; and *Euphorbia cyathophora* Murray (fire-on-the-mountain): systemic mosaic (inoculated leaves remain symptomless).

Host Range and Epidemiology

PnMV is thought to occur worldwide. It naturally infects cultivated *E. pulcherrima* and *E. fulgens* Karw. ex Klotzsch and is experimentally transmissible to eight additional *Euphorbia* spp. and to *Nicotiana* spp. There are no local lesion hosts known, and no vectors have been identified. In one study, the virus was not transmitted via standard production handling practices or by leaf or root contact with infected plants over a period of several months.

Management

Cell suspension culture has been used to eliminate both PnMV and poinsettia cryptic virus from *E. pulcherrima*. The cell aggregates were regenerated into rooted plants on solid medium. A high percentage of the plantlets thus generated were free of both viruses. This technique was noted to be inappropriate for chimeral cultivars, which will not regenerate uniform plantlets. Heat therapy has also been effective in generating virus-free minicuttings (1–1.5 cm) and meristems (0.25 mm). Poinsettia mosaic-free stock production should be possible in situations where workers are not also handling virus-infected poinsettia material, which creates the possibility of recontamination through sap transmission.

Selected References

Chiko, A. W. 1983. Poinsettia mosaic virus in British Columbia. Plant Dis. 67:427-428.

Fulton, R. W., and Fulton, J. L. 1980. Characterization of a tymo-like virus common in poinsettia. Phytopathology 70:321-324.

Koenig, R., and Lesemann, D.-E. 1980. Two isometric viruses in poinsettias. Plant Dis. 64:782-784.

Koenig, R., Lesemann, D.-E., and Fulton, R. W. 1986. Poinsettia mosaic virus. Descriptions of Plant Viruses, no. 311. Commonwealth Mycological Institute and Association of Applied Biologists, Kew, England.

Lesemann, D.-E., Koenig, R., Huth, W., Brunt, A. A., Philips, S., and Barton, A. J. 1983. Poinsettia mosaic virus—A tymovirus? Phytopathol. Z. 107:250-262.

Paludan, N., and Begtrup, J. 1986. Inactivation of poinsettia mosaic virus and poinsettia cryptic virus in *Euphorbia pulcherrima* using heat-treated mini-cuttings and meristem-tip culture. Tidsskr. Planteavl 90:283-290.

Pfannenstiel, M. A., Mintz, K. P., and Fulton, R. W. 1982. Evaluation of heat therapy of poinsettia mosaic and characterization of the viral components. Phytopathology 72:252-254.

Preil, W., Koenig, R., Engelhardt, M., and Meier-Dinkel, A. 1982. Eliminierung von Poinsettia mosaic virus (PoiMV) und Poinsettia cryptic virus (PoiCV) aus *Euphorbia pulcherrima* Willd. durch Zellsuspensionskultur. Phytopathol. Z. 105:193-197.

Tobacco Mosaic

Tobacco mosaic tobamovirus (TMV) has a worldwide distribution and affects a wide range of ornamentals. It is particularly infectious because it forms stable crystals that are easily transmitted when plant material is handled.

Symptoms

Cyclamen (*Cyclamen persicum* Mill.) infected with both TMV and potato virus X show curled, deformed petals with necrotic streaks. Plants infected with TMV plus tomato aspermy virus also show deformed flowers, whereas cyclamen infected with TMV alone show only mild symptoms of stunting and chlorosis but no flower effects. In lisianthus (*Eustoma grandiflorum* (Raf.) Shinn.), TMV produces a mosaic. Combined infections with TMV and either cucumber mosaic virus or bean yellow mosaic virus cause more severe symptoms in lisianthus, including necrotic spotting and stunting. Infection in *Gerbera jamesonii* H. Bolus ex Adlam (African daisy) has been reported from Turkey and in *Pelargonium* spp. (geraniums) from Belgium. In New Guinea impatiens, TMV causes stunting and leaf distortion (Plate 177).

TMV rarely causes economic damage to cultivated gesneriads because the effects are generally inconspicuous. Downward cupping of leaves and elongated shoots have been observed in gloxinia (*Sinningia speciosa* (Lodd.) Hiern) infected with TMV. TMV has been found in three cultivars of *Achimenes*: Johanna Michelsen, which exhibits leaf mottle and stunted, narrow, malformed leaves, and Paul Arnold and Tarantella, which are asymptomatic.

Causal Agent

TMV, the type member for the genus *Tombamovirus,* has rod-shaped particles that measure 300 × 18 nm and is a very stable virus.

Host Range and Epidemiology

Plants may become infected from root contact with virus particles or bits of infected plant tissue. Transmission of TMV by splashing during normal overhead watering has been shown. Although it is easily transmitted through handling of plant material, TMV is not known to be commonly vectored by arthropods. The type strain of TMV is not spread through seed or pollen, but other tobamoviruses are sometimes carried on seed. TMV is of particular concern in orchid culture, where a strain of the virus known as Odontoglossum ringspot virus causes severe symptoms. In other greenhouse flower crops, TMV may produce symptoms by itself, or it may aggravate symptoms on plants with multiple viral infections. It is known to infect *Bougainvillea spectabilis* Willd., *C. persicum*, *G. jamesonii*, gesneriads, New Guinea impatiens, *E. grandiflorum*, *Pelargonium* spp., and *Solanum* spp. Other flowering potted plants are likely to be additional hosts.

In a survey of gesneriads, TMV was detected frequently in cultivated plants from four states. No TMV was detected from herbarium specimens of the same genera collected from the wild, indicating that the source of TMV for the cultivated plants was quite possibly contaminated tobacco products and that the virus had presumably been spread by vegetative propagation. The virus was detected frequently (31–89%) in *Achimenes, Codonanthe, Columnea, Kohleria,* and *Smithiantha* spp. and more rarely (2–8%) in *Aeschynanthus, Episcia,* and *Nematanthus* spp. Only *Achimenes* was found to be affected with the U1 strain; all other plants were predominantly infected with the U2 strain. Both U1 and U2 strains are common in tobacco. Although 16 *Saintpaulia ionantha* Wendl. (African violet) specimens were examined, none was found to be infected with TMV.

Diagnostic species—*Nicotiana tabacum* L. (tobacco) cultivars Turkish, Turkish Samsun, Samsun, White Burley, Burley, and Xanthi: faint chlorotic lesions on the inoculated leaves, veinclearing followed by mosaic on systemically infected young leaves, and possibly distortion and blistering; *N. tabacum* cultivars Samsun NN and Xanthi-nc, *Phaseolus vulgaris* L. cv. Pinto (pinto bean), and *Chenopodium amaranticolor* Coste & Reyn. (pigweed): necrotic local lesions at temperatures below 28°C; and *N. sylvestris* Speg. & Comes and *N. tabacum* cv. Java: systemic infection. Inclusion bodies associated with TMV and easily detected by light microscopy include hexagonal crystalline plates; lateral aggregates; linear aggregates of needles, spindles, and fibers; and amorphous bodies.

Management

Workers should wash their hands after handling tobacco products, and smoking should not be allowed in greenhouses. The virus is very stable outside plants and is easily mechanically transmitted. Weeds should be eliminated from the greenhouse floor.

Selected References

Allen, W. R. 1975. Tobacco mosaic virus (type strain). Descriptions of Plant Viruses, no. 151. Commonwealth Mycological Institute and Association of Applied Biologists, Kew, England.

Allen, W. R. 1981. Dissemination of tobacco mosaic virus from soil to plant leaves under greenhouse conditions. Can. J. Plant Pathol. 3:163-168.

Gera, A., and Cohen, J. 1990. The natural occurrence of bean yellow mosaic, cucumber mosaic and tobacco mosaic viruses in lisianthus in Israel. Plant Pathol. 39:561-564.

Kaminska, M. 1975. Susceptibility of cyclamen to certain viruses. Pr. Inst. Sadow. Skierniewicach, Ser. B 1:167-172.

Yorganici, U., and Karaca, I. 1974. Tobacco mosaic virus on *Gerbera jamesonii* in Turkey. J. Turk. Phytopathol. 3:116-123.

Zaitlin, M. 1977. Tobamovirus group. Descriptions of Plant Viruses, no. 184. Commonwealth Mycological Institute and Association of Applied Biologists, Kew, England.

Zettler, F. W., and Nagel, J. 1983. Infection of cultivated gesneriads by two strains of tobacco mosaic virus. Plant Dis. 67:1123-1125.

Tobacco Ringspot of *Clerodendrum, Dahlia,* Geranium, *Hydrangea, Sinningia, Begonia,* and *Impatiens*

Tobacco ringspot nepovirus (TRSV) is an important virus that affects a wide range of annual and perennial crops in North America and, occasionally, greenhouse ornamentals. The name pelargonium ringspot virus has been used to refer to both TRSV and tomato ringspot virus on geraniums (*Pelargonium* spp.).

Symptoms

TRSV-infected impatiens (*Impatiens wallerana* Hook. f.) are stunted and show mosaic and chlorotic ring spots and line patterns. On *Begonia*, TRSV causes begonia yellow spot. Fibrous rooted begonias are stunted and develop pale yellow or white ring spots or a pale mottle. The symptoms of TRSV in begonia are similar to genetic variegation. Broad bean wilt virus causes similar symptoms in begonia. TRSV has also been identified in *Dahlia, Hydrangea,* and *Sinningia* spp.

Florist's geraniums (*P.* × *hortorum* L. H. Bailey) inoculated with TRSV as young seedlings develop chlorotic rings and line patterns that fade as the plants mature. Older geraniums are not susceptible to mechanical inoculation with TRSV. Symptomless infected *Pelargonium* plants carry a low titer of virus.

Causal Agent

TRSV is a member of the genus *Nepovirus*, which includes viruses having three isometric particles and a bipartite genome of single-stranded RNA. Nepoviruses cause ring spots and mottling on a wide range of plants. Transmission is by nematodes or pollen and, experimentally, by sap inoculation. Seed transmission may occur with nepoviruses, and a high percentage of the seed may carry the virus. Latent infections of seedlings grown from virus-infected seed have been noted.

TRSV has three types of spherical particles with angular outlines, all of them about 28 nm in diameter. Two of the particle types contain the RNA genome; the third contains solely protein. Satellite RNA is associated with some isolates of TRSV. Individual virus particles are hard to distinguish in electron microscope views because they resemble ribosomes, but the tubular or crystalline inclusions facilitate virus recognition. Serological tests can distinguish TRSV from tomato ringspot virus and other nepoviruses.

Host Range and Epidemiology

TRSV is endemic in eastern and central North America and has been occasionally reported from other parts of the world, presumably resulting from the movement of infected nursery stock. Natural hosts of TRSV include both woody and herbaceous species. Nematodes in the genus *Xiphinema* are the most well-known vectors of TRSV, but other vectors may be involved as well. Onion thrips (*Thrips tabaci* Lindeman), mites (*Tetranychus*), grasshoppers (*Melanoplus*), tobacco flea beetles (*Epitrix hirtipennis* (Melsheimer)), and aphid species may all serve as vectors in some circumstances. Seed transmission is believed to be possible in all hosts and occurs in up to 100% of the seed, where the virus is associated with the embryo. Seed transmission has not been demonstrated for *P.* × *hortorum*.

Diagnostic species—*Chenopodium amaranticolor* Coste & Reyn. (pigweed) and *C. quinoa* Willd. (quinoa): necrotic local lesions (the lack of systemic symptoms on *C. quinoa* distinguishes TRSV from tomato ringspot virus); *Cucumis sativus* L. (cucumber): chlorotic or necrotic local lesions and systemic mottle and stunting with distortion of the growing point; *Nicotiana tabacum* L. and *N. clevelandii* Gray. (tobacco): necrotic local lesions that may turn into rings and systemic ring or line patterns; *Phaseolus vulgaris* L. (common bean): necrotic local lesions and systemic spots, rings, and tip necrosis; *Lycopersicon esculentum* Mill. (tomato) (difficult to inoculate): small, necrotic, local lesions and systemic vein necrosis and epinasty; and *Vigna unguiculata* (L.) Walp. subsp. *unguiculata* (black-eyed pea): necrotic local lesions, systemic necrosis, and wilt.

Management

Virus-indexing programs will eliminate TRSV from production. Contamination of growing media by nematodes should be avoided through the use of soilless media or pasteurized or fumigated soil.

Selected References

Abo El-Nil, E., Hildebrandt, A. C., and Evert, R. F. 1976. Symptoms induced on virus-free geranium seedlings by tobacco ringspot and cucumber mosaic viruses. Phyton (Horn, Austria) 34:61-64.

Khan, M. A., and Maxwell, D. P. 1975. Identification of tobacco ringspot virus in *Clerodendrum thomsoniae*. Phytopathology 65:1150-1153.

Lockhart, B. E., and Betzhold, J. A. 1979. Begonia yellow spot disease caused by tobacco ringspot virus infection. Plant Dis. Rep. 63:1046-1047.

Lockhart, B. E., and Pfleger, F. L. 1979. Identification of a strain of tobacco ringspot virus causing a disease of impatiens in commercial greenhouses. Plant Dis. Rep. 63:258-261.

Romaine, C. P., Newhart, S. R., and Anzola, D. 1981. Enzyme-linked immunosorbent assay for plant viruses in intact leaf tissue disks. Phytopathology 71:308-312.

Scarborough, B. A., and Smith, S. H. 1977. Effects of tobacco and tomato ringspot viruses on the reproductive tissues of *Pelargonium* × *hortorum*. Phytopathology 67:292-297.

Stace-Smith, R. 1985 Tobacco ringspot virus. Descriptions of Plant Viruses, no. 309. Commonwealth Mycological Institute and Association of Applied Biologists, Kew, England.

Walsh, D. M., Horst, R. K, and Smith, S. H. 1974. Factors influencing symptom expression and recovery of tobacco ringspot virus from geranium. (Abstr.) Phytopathology 64:588.

Tobacco Streak
of *Dahlia* and *Impatiens*

Symptoms

In *Dahlia* hybrids, tobacco streak ilarvirus (TSV) may be asymptomatic or cause mottling. A strain of TSV has been noted on *Impatiens* sp., causing stunting of the whole plant and twisting and deformation of leaves. Flowers are reduced in number but are not deformed.

Causal Agent

TSV is a member of the genus *Ilarvirus*, whose members have a three-part RNA genome in three separate isometric particles containing single-stranded RNA. Particle size varies from 26 to 35 nm. The ilarviruses tend to have a wide host range and are easily transmitted by sap. Characteristics of inclusion bodies have not been defined for these viruses. Thrips, seed, and pollen transmission are known for ilarviruses. TSV has three isometric particles, 27, 30, and 35 nm in diameter.

Host Range and Epidemiology

TSV is widespread and has a broad host range, including species in more than 30 families of monocots and dicots. Certain thrips, including the western flower thrips, *Frankliniella occidentalis* (Pergande), and the onion thrips, *Thrips tabaci* Lindeman, have been shown to transmit TSV. Transmission (at a high percentage in some cases) through seed and pollen is known for several hosts. Pollen transmission of the virus may be to the pollinated plant as well as to the resulting seed. In one instance, thrips were reported to facilitate virus spread by feeding on field crop leaves that were dusted with windborne pollen from adjacent virus-infected weeds.

Diagnostic species—*Nicotiana tabacum* L. (tobacco): necrotic local spots or rings and systemic line and oak leaf patterns followed by recovery; *Cyamopsis tetragonoloba* (L.) Taub. (cluster bean): necrotic local lesions; and *Vigna unguiculata* (L.) Walp. subsp. *cylindrica* (catjang): chlorotic or reddish, necrotic local lesions followed by systemic necrosis or mottling.

Symptoms may vary with virus strain, so identification by serological techniques is preferred. Virus particles occur in aggregates in young leaflets and apical tissues.

Management

Control of weed hosts and thrips vectors, particularly during hybrid seed production, is critical. Because of its wide host range, TSV may become a significant threat to greenhouse flower production in cases in which the virus is introduced and there is a significant population of thrips.

Selected References

Fulton, R. W. 1985. Tobacco streak virus. Descriptions of Plant Viruses, no. 307. Commonwealth Mycological Institute and Association of Applied Biologists, Kew, England.

Lockhart, B. E., and Betzold, J. A. 1980. Leaf-curl of impatiens caused by tobacco streak virus infection. Plant Dis. 64:289-290.

Tomato Bushy Stunt
of Geranium and *Primula*

A disease called geranium crinkle was first described from Germany in 1927 and from the United States in 1940. The viral agent is now often referred to as pelargonium leaf curl virus (PLCV). A pelargonium strain of tomato bushy stunt tombusvirus (TBSV) is responsible for these diseases on geranium. Hosts of TBSV also include *Primula* spp. (primroses).

Symptoms

Florist's geraniums (*Pelargonium* × *hortorum* L. H. Bailey) show TBSV symptoms clearly; yellow, stellate spots on younger leaves become visible during late winter. The centers of these spots usually become necrotic, and as the surrounding tissue expands, the leaves buckle and become distorted. This malformation is referred to as crinkling. Shot holing may develop on severely affected leaves. Elongated, brown, necrotic spots may develop on petioles and stems, and the entire top of the plant may brown and die. The plant will often outgrow the symptoms during the hot summer months. The virus cannot be isolated from geraniums during the summer, although it can be experimentally sap transmitted between all other hosts at all times. A different strain of TBSV causes a systemic necrosis in German primrose, *P. obconica* Hance.

Causal Agent

TBSV is in the genus *Tombusvirus*. The members of this group have a single-stranded RNA genome that is monopartite and isometric particles approximately 30 nm in diameter. Tombusviruses have a wide host range. Infection has been shown to occur through virus particles in the soil for some tombusviruses; sap inoculation allows artificial transmission. Temperature and photoperiod are thought to be important factors in symptom development in tombusvirus infections of tomatoes (*Lycopersicon esculentum* Mill.) and other plants. TBSV from geranium has spherical particles about 28 nm in diameter.

Host Range and Epidemiology

The virus does not appear to be transmissible in geraniums (*Pelargonium* spp.) by handling, by plant leaf or root contact, or by cutting knife. There are no indications of transmissibility to geraniums via seed, soil, or insects. Experimentally, the TBSV strain from geranium will infect at least 54 plant species.

Diagnostic species—*Phaseolus vulgaris* L. (common bean): pale brown, necrotic local lesions and stellate lesions followed by systemic mottle; *Datura stramonium* L. (jimsonweed): chlorotic local lesions that become necrotic followed by systemic infection; and *Chenopodium amaranticolor* Coste & Reyn. (pigweed): systemic infection.

Management

Heat treatment at 37°C for 4 weeks has been reported effective for freeing plants of TBSV. TBSV's characteristically slow spread suggests that it may not be a serious threat to commercial cultivation of geraniums. The disease persists in commercial crops, however, in spite of roguing. Normal cultural operations cause little or no spread. Once clean stock is established through heat treatment, reintroduction of this virus is unlikely in the course of geranium production.

Selected References

Bawden, F. C., and Kassanis, B. 1947. *Primula obconica*, a carrier of tobacco necrosis virus. Ann. Appl. Biol. 34:127-135.

Hollings, M. 1962. Studies of Pelargonium leaf curl virus. I. Host range, transmission and properties in vitro. Ann. Appl. Biol. 50:189-202.

Jones, L. K. 1940. Leaf curl and mosaic of geranium. Pages 3-19 in: Wash. Agric. Exp. Stn. Bull. 390.

Martelli, G. P., Quacquarelli, A., and Russo, M. 1971. Tomato bushy stunt virus. Descriptions of Plant Viruses, no. 69. Commonwealth Mycological Institute and Association of Applied Biologists, Kew, England.

Reinert, R. A. 1962. Symptoms, transmission and control of virus diseases of *Pelargonium*. Ph.D. thesis. University of Wisconsin, Madison.

Tomato Ringspot
of Geranium and *Hydrangea*

Symptoms

On hydrangea (*Hydrangea macrophylla* (Thunb.) Ser.), tomato ringspot nepovirus (ToRSV) causes stunting of the plant, leaves, and flowers; occasional leaf distortion; and a dull yellow leaf chlorosis marked with dark green blotches. The flowers may open erratically and have a mixture of green and colored flowers in the same flower head. On the florist's geranium (*Pelargonium* × *hortorum* L. H. Bailey), ToRSV causes chlorotic rings in older leaves that are visible in the spring. It may also have an important role in enhancing symptoms caused by other viruses. Geraniums infected with both ToRSV and pelargonium flower break virus may show symptoms over a longer portion of the year than those infected with ToRSV alone.

Causal Agent

ToRSV is in the genus *Nepovirus* (see the section on tobacco ringspot virus for a description of nepoviruses). ToRSV has three, 28-nm particles with angular outlines, two of which carry its single-stranded RNA genome. Serological tests are used to distinguish ToRSV from tobacco ringspot virus and other closely related nepoviruses.

Host Range and Epidemiology

The virus has a wide experimental and natural host range, which includes woody and herbaceous plants. *Pericallis* × *hybrida* (cineraria) has been systemically infected experimentally. Nematodes in the genus *Xiphinema* are responsible for transmitting ToRSV in the field. In one test, researchers were unable to transmit ToRSV to healthy *H. macrophylla* cuttings with a knife previously used on a virus-infected plant. The virus is seed transmitted in *P.* × *hortorum*. Infected geranium seedlings may be stunted but show no further symptom development. When mature geraniums are infected, the virus is present at low titers, making detection difficult.

Perennial weed hosts are important for both the survival and spread of ToRSV. Windborne seeds of *Taraxacum officinale* F. H. Wigg. (dandelion), for example, may spread the virus over long distances. ToRSV appears to be endemic in North America and has been reported to be common on *Pelargonium* spp. in some European countries.

Diagnostic species—*Chenopodium amaranticolor* Coste & Reyn. (pigweed) and *C. quinoa* Willd. (quinoa): small chlorotic or necrotic local lesions and systemic tip dieback (systemic symptoms on *C. quinoa* distinguish this virus from the closely related tobacco ringspot virus); *Cucumis sativus* L. (cucumber): local chlorotic spotting, systemic chlorosis, and mottle; *Lycopersicon esculentum* Mill. (tomato): local necrotic flecks, systemic necrosis, and mottle; *Nicotiana tabacum* L. (tobacco): necrotic local lesions or rings and systemic necrotic or chlorotic ring and line patterns; *Petunia* × *hybrida* Vilm.: necrotic local lesions and collapse of young, systemically infected leaves; *Phaseolus vulgaris* L. (common bean): chlorotic local lesions, systemic rugosity, and necrosis of young leaves; and *Vigna unguiculata* (L.) Walp. (cowpea): chlorotic or necrotic local lesions and systemic tip necrosis.

Management

Soilless growing medium or treated field soil should be used to avoid nematodes. Stock plants should be virus-indexed for this and other viruses. Weeds should be controlled to eliminate sources of virus.

Selected References

Brierly, P. 1954. Symptoms in the florists' hydrangea caused by tomato ringspot virus and an unidentified sap-transmissible virus. Phytopathology 44:696-699.

Christensen, O. V., and Paludan, N. 1978. Growth and flowering of pelargonium infected with tomato ringspot virus. J. Hortic. Sci. 53:209-213.

Kemp, W. G. 1969. Detection of tomato ringspot virus in *Pelargonium* in Ontario. Can. Plant Dis. Surv. 49:1-4.

Paludan, N., and Begtrup, J. 1987. *Pelargonium × hortorum.* Pelargonium flower break virus and tomato ringspot virus: Infection trials, symptomatology and diagnosis. Dan. J. Plant Soil Sci. 91:183-194.

Reinert, R. A., Hildebrandt, A. C., and Beck, G. E. 1963. Differentiation of viruses transmitted from *Pelargonium hortorum.* Phytopathology 53.1291-1298.

Ryden, K. 1972. Pelargonium ringspot—A virus disease caused by tomato ringspot virus in Sweden. Phytopathol. Z. 73:178-182.

Scarborough, B. A., and Smith, S. H. 1977. Effects of tobacco and tomato ringspot viruses on the reproductive tissues of *Pelargonium × hortorum.* Phytopathology 67:292-297.

Stace-Smith, R. 1984. Tomato ringspot virus. Descriptions of Plant Viruses, no. 290. Commonwealth Mycological Institute and Association of Applied Biologists, Kew, England.

Stone, O. M. 1980. Nine viruses isolated from pelargonium in the United Kingdom. Acta Hortic. 110:113-122.

Tomato Spotted Wilt
and Impatiens Necrotic Spot

The genus *Tospovirus* is composed of a few viruses that until recently were all considered to be strains of tomato spotted wilt virus (TSWV). TSWV had long been considered unique among plant viruses and had been placed in a group by itself. Recent advances in our understanding have led to the classification of TSWV within the family Bunyaviridae. Members of this virus family cause diseases of mammals and plants as well as of their insect vectors. The virus particles of tosposviruses are quasispherical and unique in that they are enveloped by a membrane composed of lipid and protein.

During the mid-1980s, TSWV became more widely distributed in the United States as the range of the vector, the western flower thrips (*Frankliniella occidentalis* (Pergande)) (Fig. 51), extended eastward across the continent. At the same time, the western flower thrips began to be a serious greenhouse pest, and what was thought to be TSWV became much more frequently encountered in greenhouses across the United States and Canada. The thrips and the virus quickly became distributed throughout the North American greenhouse industry because of the frequent exchange of plants and cuttings across long distances. The European greenhouse industry also began experiencing outbreaks of TSWV during the late 1980s, coinciding with the importation of the western flower thrips.

In 1989, a serologically distinct TSWV strain was reported from *Impatiens* and designated TSWV-I to distinguish it from the common or lettuce strain, TSWV-L. The use of differentiating antisera allowed data to be collected on the relative incidence of the two strains, and TSWV-I was found more frequently in American flower industry greenhouses. In 1991, TSWV-I was more thoroughly characterized and was determined to be a distinct virus, at which time it was named impatiens necrotic spot virus (INSV).

INSV is a serious threat to a wide range of North American greenhouse flower crops, especially *Impatiens wallerana* Hook. f., New Guinea impatiens, *Cyclamen persicum* Mill., *Pericallis × hybrida* R. Nordenstam (cineraria), *Sinningia speciosa* (Lodd.) Hiern (gloxinia), and *Exacum affine* Balf. f. (Persian violet). In a Pennsylvania survey of greenhouse potted plants undertaken from December 1989 to February 1991, INSV was detected in 247 tospovirus-infected plants, TSWV in only 25, and mixed infections of the two viruses in six of the plants. Additionally, 95% of the tospovirus-infected bedding plants and 96% of the tospovirus-infected perennials were infected with INSV. In North American greenhouses, TSWV is detected in *Dahlia* hybrids and chrysanthemum (*Dendranthema × grandiflorum* Kitam.) more often than in other flower crops. Flowering annuals are more often infected with INSV than with TSWV. Both viruses are threats to vegetable transplants. In Europe, TSWV is the predominant tospovirus in *Lycopersicon esculentum* Mill. (tomato), *Capsicum annuum* L. (pepper), *Gerbera jamesonii* H. Bolus ex Adlam (African daisy), *D. × grandiflorum,* and *Dieffenbachia maculata* (Lodd.) G. Don grown in greenhouses. INSV is reported to be less widespread in Europe than in the United States, but it has been identified in *Begonia* in the Netherlands and in *Anemone coronaria* L. in Germany.

Symptoms

TSWV and INSV cause an extraordinarily broad range of symptoms on many plants (Plates 178–217). The possible symptoms include stunting, necrotic spotting, chlorotic spotting, areas of black or brown stem necrosis, ring spots, mosaic, line patterns, and vein necrosis. Plants infected while young are the most severely affected. Complete necrosis and collapse of seedlings has been observed, especially in the case of young gloxinias infected with INSV. Any of the plant hosts may exhibit several of the symptom types, showing variation from cultivar to cultivar and from plant to plant. Latent (symptomless) infections are common in certain cultivars of susceptible crops. Environmental conditions also affect symptom expression. Previously symptomless *I. wallerana* and New Guinea hybrids, for example, may show symptoms during the winter months. Symptoms of INSV and TSWV may be indistinguishable on a given host.

Fig. 51. Male (left) and female (right) western flower thrips (*Frankliniella occidentalis*), the most common greenhouse vector of tospoviruses. (Courtesy M. P. Parrella)

Rieger begonias (*Begonia × hiemalis* Fotsch.) infected with INSV show an overall yellow mottling of the leaf (Plate 188); large, round, brown necrotic spots; or browning of the veins at the petiole end of the leaf (Plate 189). Petioles may turn brown and collapse in wax begonia (*Begonia* Semperflorens-Cultorum hybrids), or an area of the leaf blade at the petiole end may become necrotic. Zigzag line patterns may appear in the leaves.

Catharanthus roseus (L.) G. Don (vinca) is used as a diagnostic species for TSWV. Local black spots form 10–14 days after inoculation, and leaves may yellow and abscise. Eventually, mosaic and leaf deformation develop. Leaf deformity has also been observed in INSV infection.

Seed transmission of TSWV is reported for *P. × hybrida*. Leaves of cinerarias infected with INSV show small, round, yellow spots; larger chlorotic patches with diffuse borders (Plate 190); necrotic patches; or brown, white, or yellow ring spots. Petioles may turn purple or blacken (Plate 191), and lower leaves may wilt. Flowers may bloom fairly normally on cinerarias with devastated foliage. Symptoms of INSV in calceolaria (*Calceolaria* Herbeohybrida group) are quite similar; plants may be stunted (Plates 192 and 193).

Cyclamen infected with INSV often show round, brown, necrotic spots (Plate 194). Brown or yellow ring spots may also appear (Plate 195). Browning at the petiole end of the leaf may occur, but this is not a distinguishing trait, since similar symptoms may develop on plants with bacterial soft rot or Fusarium wilt. Brown, zigzag line patterns have been observed, as has flower distortion and petal breaking (Plate 196). Symptoms may develop months after the initial infection.

Dahlias (*Dahlia* hybrids) produced from tubers are frequently affected by TSWV. Yellow line patterns in the leaves are typical (Plate 197), as are yellow ring spots or blotches (Plate 198). Dahlias produced from seed show light chlorotic spots or line patterns when infected with INSV.

Exacum with INSV are hard to distinguish from those injured by *Botrytis cinerea*. Sections of branches turn light tan and collapse, and associated leaves wither. Black, discolored, cankerlike areas may appear on the main stem (Plate 199). Infection by *Botrytis* sp. may follow virus injury and obscure the diagnosis.

In hydrangea (*Hydrangea macrophylla* (Thunb.) Ser.), TSWV symptoms are not unique because TSWV shares with hydrangea ringspot virus the ability to cause ring spots on this host.

On New Guinea impatiens, symptoms of INSV include stunting; brown or purple leaf spots; ring spots on petals; black, diffuse spotting on leaves; black stem sections; leaf stunting; distortion; and chlorotic mottling (Plates 200–203). Plants may wilt and collapse. Variation is extreme from cultivar to cultivar, and many cultivars are symptomless. Symp-

toms appear during the winter and may be masked during the summer. Symptoms caused by TSWV are distinguishable only by enzyme-linked immunosorbent assay (ELISA).

Symptoms of INSV on single- and double-flowered *I. wallerana* include yellow mosaic; scattered, small, black or brown spots or flecks; brown or black ring spots often preceding leaf yellowing and abscission; black stem sections (Fig. 52); and leaf and plant stunting. Double-flowered impatiens (which are most often vegetatively propagated) are highly attractive to the thrips vector and often show dramatic symptoms of INSV infection (Plate 204). TSWV infection is less common, and symptoms are similar.

Yellow spots and ring spots on *Pelargonium peltatum* (L.) L'Hér. (ivy geranium) have been associated with the presence of TSWV (Plate 206). There have also been a few reports of INSV and TSWV in zonal geraniums with viruslike symptoms. Symptomless infections are known to occur on *P. × hortorum* L. H. Bailey (florist's geranium). Symptoms have been produced on *P. × domesticum* L. H. Bailey (regal geranium) by inoculation.

The fairy primrose (*Primula malacoides* Franch.) shows stunting, yellowing, and withering of leaves when infected with TSWV. TSWV is usually fatal in the German primrose (*P. obconica* Hance), whose leaves develop marginal chlorosis, necrotic spotting, and wilting. Stunting, necrotic veins, irregular necrotic lesions, and mottling may appear on Chinese primrose (*P. sinensis* Sabine ex Lindl.) infected with INSV (Plate 207).

INSV-infected ranunculus (*Ranunculus asiaticus* L.) may show foliar necrosis. Flower buds may become necrotic (Plate 208), and the entire plant may be stunted (Plate 209).

Gloxinia is one of the most severely injured INSV hosts. Young plants may brown from the center, wilt, and die (Plate 210). Chlorotic or necrotic brown line patterns in zigzags, brown spots, or ring spots may appear in the foliage of plants that acquired the virus at a more mature stage of growth (Plates 210–214). Young leaves may also exhibit symptoms of systemic infection (Plate 215). Local lesions may appear as soon as 2 days after thrips feeding on this host (Plate 216).

Causal Agents

TSWV and INSV have roughly spherical particles, 70–120 nm in diameter, which are enveloped in a protein-studded membrane (Fig. 53). There are four structural proteins contained in the viral particle: two glycoproteins (G1 and G2) located in the membrane; a nucleoprotein (N protein), which surrounds the RNA; and an enzyme, which synthesizes new RNA. The genetic information of the viruses is contained in

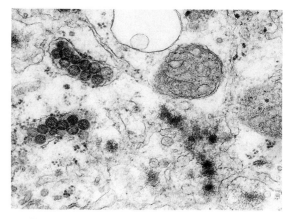

Fig. 53. Electron micrograph of impatiens necrotic spot tospovirus (INSV) within a tobacco cell. Note two groups of mature, single-enveloped INSV virions in swollen cisternae of rough endoplasmic reticulum (left) and dark nucleocapsid aggregates of INSV (right). The large, round, gray structure is a tobacco mitochondrion. (Courtesy R. Lawson)

Fig. 52. Blackened stem sections are one of the most characteristic symptoms of impatiens necrotic spot virus infection on impatiens.

70

three single-stranded RNA molecules. The two tospoviruses, TSWV and INSV, although distinguished by serologically distinct N protein, cause similar symptoms and have overlapping host ranges.

The two viruses may be distinguished by ELISA or other tests that employ serological techniques. TSWV and INSV have different N protein components. They also differ in the nature of the inclusion bodies formed in the cytoplasm of diseased tissues. INSV typically forms viroplasms containing protein filaments in a paracrystalline array, while TSWV isolates form loose bundles of filaments (Fig. 53 and Plate 168). There is significant variation among isolates of both viruses in the abundance of inclusion bodies and virions.

New virus forms may be detected in the future that are not currently detectable with available antisera. The U.S. Department of Agriculture Florist and Nursery Crops Laboratory reported that an INSV isolate from gloxinia that had been maintained at a high temperature in *Nicotiana* (tobacco) no longer reacted with INSV antisera because of the lack of the nucleoprotein accumulations in viroplasms typical of INSV. Bioassays of inoculated plants such as *N. benthamiana* Domin are valuable for diagnosis of variants from the TSWV and INSV types currently known (those that do not test positive by ELISA).

Host Range and Epidemiology

TSWV is considered common in subtropical and tropical regions of the world and, together with INSV, has recently become more important in the Northern Hemisphere because of the increased range of one of its vectors, *F. occidentalis*. TSWV has an extremely broad experimental host range, encompassing nearly 500 species in 50 plant families, including monocots and dicots. INSV also appears to have a wide host range.

Many flower crops are hosts of TSWV and INSV, as are many vegetables and weeds. In addition to the hosts listed above, other flowering potted plant hosts for tospoviruses include *Anemone, Aquilegia, Browallia, Calceolaria, Campanula, Capsicum, Clerodendrum, Eustoma, Fuchsia, Gardenia, Gerbera, Kalanchoe, Lantana, Mimulus, Ranunculus, Saintpaulia, Schlumbergera, Solanum, Schizanthus,* and *Streptocarpus.* The geographic range and host range of INSV is as yet undetermined, but it appears that many greenhouse flower crops known to be susceptible to TSWV are also susceptible to INSV. Double infections of the two viruses are occasionally detected. Both INSV and TSWV cause disease losses in American and European floricultural production.

Diagnostic species—*Petunia* × *hybrida* Vilm. 'Pink Beauty' and 'Minstrel': necrotic local lesions; *N. tabacum* L. 'Samsun NN,' *N. clevelandii* Gray., and *N. glutinosa* L. (tobacco): necrotic local lesions followed by leaf deformation and systemic necrotic patterns; *Cucumis sativus* L. (cucumber): chlorotic local lesions with necrotic centers on cotyledons; *C. roseus:* black local lesions, sometimes chlorosis and abscission of leaves, systemic mosaic, and deformation; and *Tropaeolum majus* L. (nasturtium): inoculated leaves symptomless; systemic mosaic of yellow and dark green specks develops in 8–12 days, sometimes with necrotic spots.

TSWV is vectored only by certain species of thrips: *F. fusca* (Hinds), the tobacco thrips; *F. intonsa; F. occidentalis; F. schultzei* (Trybom), the common blossom thrips; *Scirtothrips dorsalis* Hood; *Thrips palmi* Karny; *T. setosus* Moulton; and *T. tabaci* Lindeman, the onion thrips. Many of these thrips vectors have wide host ranges, facilitating spread of the virus. TSWV is acquired by larval thrips that feed on infected plant tissue. Virus acquisition takes 15–30 min of feeding. The virus is transmitted by the feeding of the adult (winged) thrips and, for TSWV, requires as little as 5 min. Although TSWV multiplies in the thrips, there is no evidence of transovarian transmission.

At this time, INSV is known to be vectored only by the western flower thrips. Thrips-virus interactions have not yet been studied for INSV.

A tospovirus and a thrips vector may be introduced into the greenhouse together on a thrips-infested, tospovirus-infected plant, or they may arrive in separate plant shipments. Thrips that have acquired TSWV or INSV during larval feeding on infected plants are able to efficiently spread the virus as adults, creating epiphytotics on other susceptible crops within the greenhouse. Thus far, greenhouse INSV outbreaks have been associated with high populations of a single vector, the western flower thrips. Infected flowering potted plant crops are an important source of virus for infection of vegetable and flower bedding plant crops.

Seed transmission of TSWV has been reported for *P.* × *hybrida* and *L. esculentum* but has not as yet been studied with INSV. TSWV is believed to be carried on the seed coat, rather than in the embryo. Most, if not all, of the virus spread in the North American greenhouse industry today appears to be by movement of plants or cuttings, rather than by seed.

Management

Control of TSWV and INSV is dependent on elimination of infected plants and control of the thrips vector. Specialist propagators of susceptible crops should work closely with reputable diagnostic laboratories in the development and maintenance of clean stock. Exclusion is an important control strategy for tospoviruses, and growers should insist on propagation material that is free of TSWV, INSV, and thrips. Diseased plants should be rogued. It may be helpful to halt production of highly susceptible cultivars or crop species until a successful virus and thrips management program is established. Crop material should not be carried over from season to season.

The western flower thrips feeds deep within vegetative or flower buds or on pollen at the back of deep-throated flowers, making it difficult to contact with insecticides. In addition, there are three stages in the life cycle that present no target for chemical control: the eggs, which are inserted into host plant tissue, and the prepupa and pupa, which are nonfeeding stages typically located in the soil of the greenhouse floor or in the growing medium. Also, resistance to many of the pesticide groups, including pyrethroids, organophosphates, and carbamates, has been documented for the western flower thrips.

Monitoring for thrips allows control programs to get under way promptly (Fig. 54) by letting the grower know when thrips populations have risen above an arbitrary, self-determined action threshold. Insecticide treatment thresholds of 10–20 thrips per yellow sticky card per week have been adopted by some North American growers. Significant losses caused by INSV are generally correlated with high thrips

Fig. 54. As part of a tospovirus control program, yellow sticky cards are suspended just above the crop level to allow monitoring for thrips in a greenhouse.

populations. Monitoring for the presence of the virus with indicator petunia (*Petunia × hybrida*) plants flagged with thrips-attractive blue or yellow (nonsticky) cards may be helpful. The cultivars Calypso, Super Blue Magic, and Summer Madness are good indicators. Brown rings will develop around thrips feeding scars if a tospovirus has been transmitted (Plate 217). Monitoring is particularly critical within groups of plants newly introduced from another greenhouse operation.

When thrips are detected at a level above the action threshold, an effective insecticide should be applied without delay. At least three treatments should be applied at 5-day intervals. All plants in the greenhouse should be treated, including weeds. The next cycle of treatment should be initiated when card counts indicate another population increase.

Biological control agents are being sought for the western flower thrips. The predators *Amblyseius* (*Neoseiulus*) *cucumeris* (Oudemans), *A.* (*Neoseiulus*) *barkeri* (Hughes), and *Orius tristicolor* (White) have shown some promise for control of thrips in greenhouse cucumbers. The mite *Hypoaspis* (*Geolaelaps*) *miles* (Berlese), entomophagous fungi such as *Verticillium lecanii* (A. Zimmerm.) Viégas, and insect-predatory nematodes are also being investigated for their ability to control various stages of the thrips life cycle.

Careful attention to which crops are grown together within the same greenhouse will reduce crop losses. Seed crops should be separated from cutting crops, and susceptible vegetables should not be grown with vegetatively propagated plants. Breaking the replication cycle of viruliferous thrips may be effective, e.g., with a crop-free, weed-free period in midsummer or with production of a nonhost such as poinsettias during the fall. In southern and western areas of the United States where the virus and thrips are endemic, it may be necessary to screen the growing area to reduce thrips incursion from outdoors. Screening may also be used to isolate one house for the production of highly susceptible cultivars, crops, or propagation blocks. Areas adjacent to growing structures should be free of weeds. Plant breeding and genetic engineering efforts are now being directed toward finding solutions to diseases caused by TSWV and INSV.

Selected References

Allen, W. R. 1992. Tomato spotted wilt virus-western flower thrips complex: An integrated approach to control. Pages 119-130 in: Proc. Conf. Insect Dis. Manage. Ornamentals, 8th. M. Daughtrey, ed. Society of American Florists, Alexandria, VA.

Allen, W. R., and Matteoni, J. A. 1988. Cyclamen ringspot: Epidemics in Ontario greenhouses caused by the tomato spotted wilt virus. Can. J. Plant Pathol. 10:41-46.

Allen, W. R., and Matteoni, J. A. 1991. Petunia as an indicator plant for use by growers to monitor for thrips carrying the tomato spotted wilt virus in greenhouses. Plant Dis. 75:78-82.

Cho, J. J., Mau, R. F. L., Mitchell, W. C., Gonsalves, D., and Yudin, L. S. 1987. Host list of plants susceptible to tomato spotted wilt virus (TSWV). Coll. Trop. Agric. Human Resour. Univ. Hawaii. Res. Ext. Ser. 078.

de Avila, A. C., deHaan, P., Kitajima, E. W., Kormelink, R., Resende, R. de O., Goldbach, R., and Peters, D. 1992. Characterization of a distinct isolate of tomato spotted wilt virus (TSWV) from *Impatiens* sp. in the Netherlands. Neth. J. Plant Pathol. 134:133-151.

de Avila, A. C., deHaan, P., Smeets, M. L. L., Resende, R. de O., Kormelink, R., Kitajima, E. W., Goldbach, R. W., and Peters, D. 1993. Distinct levels of relationships between tospovirus isolates. Arch. Virol. 128:211-227.

German, T., Ullmann, D., and Moyer, J. W. 1992. *Tospoviruses:* Diagnosis, molecular biology, phylogeny and vector relationships. Annu. Rev. Phytopathol. 30:315-348.

Hausbeck, M. K., Welliver, R. A., Derr, M. A., and Gildow, F. E. 1992. Tomato spotted wilt virus survey among greenhouse ornamentals in Pennsylvania. Plant Dis. 76:795-800.

Jones, R. K., and Moyer, J. W. 1986. Tomato spotted wilt virus in gloxinia in North Carolina. N.C. Flower Growers Bull. 30:11-13.

Jones, R. K., and Moyer, J. W. 1987. Exacum, a new host for tomato spotted wilt virus. N.C. Flower Growers Bull. 31:1-2.

Law, M. D., and Moyer, J. W. 1990. A tomato spotted wilt-like virus with a serologically-distinct N protein. J. Gen. Virol. 71:933-938.

Law, M. D., Speck, J., and Moyer, J. W. 1992. The M RNA of Impatiens necrotic spot tospovirus (Bunyaviridae) has an ambisense genomic organization. Virology 188:732-741.

Lawson, R. H. 1993. Varying properties of tospoviruses make detection, isolation difficult. Greenhouse Manager 11:139-140.

Pundt, L., Sanderson, J., and Daughtrey, M. 1992. Petunias are your tip-off for TSWV. Grower Talks 56:69-72.

Sether, D. M., and DeAngelis, J. D. 1992. Tomato spotted wilt virus host list and bibliography. Ore. State Univ. Agric. Exp. Stn. Spec. Rep. 888.

Tsu, H., and Lawson, R., eds. 1991. Virus-thrips-plant interactions of tomato spotted wilt virus. Proc. U.S. Dep. Agric. Workshop. U.S. Dep. Agric. Agric. Res. Serv. Publ. 87.

Urban, L. A., Huang, P.-Y., and Moyer, J. W. 1991. Cytoplasmic inclusions in cells infected with isolates of L and I serogroups of tomato spotted wilt virus. Phytopathology 81:525-529.

Urban, L. A., Speck, J., Moyer, J. W., and Daub, M. E. 1992. Transformation of chrysanthemum with the nucleocapsid gene of tomato spotted wilt virus. (Abstr.) Phytopathology 82:1147.

Diseases Caused by Nematodes

Nematodes are small (0.5–5 mm in length), nonsegmented roundworms that are common in soil (Fig. 55). Most nematodes feed on microscopic animal and plant life such as algae, fungi, bacteria, and other nematodes. Plant-parasitic nematodes are a specialized group, the majority of which are obligate parasites of plants. Their most important adaptation is the stylet, a retractable, hollow spear in the head region used for puncturing and withdrawing nutrients from plant cells.

Most plant-parasitic nematodes parasitize roots, but some species invade the stems and leaves. The root parasites, whose feeding may cause aboveground symptoms, are either ectoparasitic or endoparasitic. Ectoparasitic nematodes remain on the outside of plant roots and feed by probing into root hairs, other epidermal cells, or in some cases subsurface tissues.

Endoparasitic nematodes enter the root (or in some cases aboveground tissues) and are potentially more damaging than ectoparasites. Root-knot and cyst nematodes are sedentary endoparasites, remaining in one place within the roots their entire adult lives. The lesion nematodes and burrowing nematodes, in contrast, are migratory endoparasites, damaging roots by moving through tissues and feeding on cells. Extensive root lesions may result.

Nematodes cause disease in plants in several ways. First, they directly injure cells and tissues by feeding or by burrowing through tissues. Depending on the plant and the nematode involved, symptoms on roots may include inhibition of root elongation, swollen root tips, galls, lesions, and shortened, stubby roots. Root damage can alter uptake of nutrients and water, resulting in wilting and other aboveground symptoms. Nematode feeding can also provide wounds for the entry of fungi and bacteria that cause root rots and vascular wilts. Nematodes in the genus *Xiphinema* are the vectors of several destructive viruses of flowering potted plants.

The severity of nematode damage is related to the number of nematodes present and the susceptibility of the plant. Thus, damage often appears to worsen gradually. When only a few nematodes are present, damage may be easily overlooked or misdiagnosed as a cultural problem.

Sources of contamination include improperly treated field soil, flats of plants in contact with the ground (Fig. 56), and infected vegetative propagation material. Foliar and stem nematodes are typically introduced into the greenhouse on cuttings and rooted plants; soil-inhabiting nematodes may reside in corms, bulbs, rhizomes, roots, and field soil. Nematodes are most damaging when greenhouse crops are infected early in the production cycle. Because of the widespread use of soilless media, plant-parasitic nematodes are encountered less often in potted plants today than in the past when field soil was commonly used.

The most important plant-pathogenic nematodes of potted plants are root-knot (*Meloidogyne*), foliar (*Aphelenchoides*), and bulb and stem (*Ditylenchus*) nematodes. The life cycles and management practices for these are discussed below. A list of nematodes reported on the hosts in this compendium is given in Table 18.

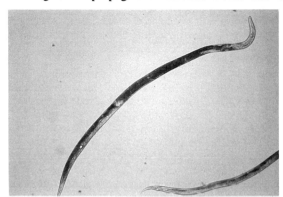

Fig. 55. *Aphelenchoides* sp., a foliar nematode. (Courtesy J. LaMondia)

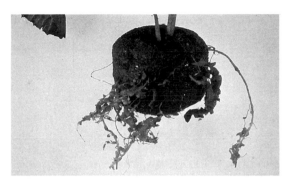

Fig. 56. Root knot developing on roots that grew out into nematode-infested soil on a greenhouse floor.

TABLE 18. Nematodes Reported on Flowering Potted Plants

Host	Nematodes
Aquilegia sp. (columbine)	*Aphelenchoides* sp., *A. ritzemabosi*, *Meloidogyne incognita acrita*
Begonia sp.	*Aphelenchoides fragariae*, *Ditylenchus destructor*, *Helicotylenchus* sp., *Hoplolaimus* sp., *Meloidogyne arenaria*, *M. incognita*, *M. incognita acrita*, *M. hapla*, *M. javanica*, *Pratylenchus penetrans*, *Rotylenchus reniformis*
Capsicum annuum (ornamental pepper)	*Belonolaimus gracilis*, *Dolichodorus heterocephalus*, *Helicotylenchus dihystera*, *Hemicycliophora arenaria*, *Meloidogyne hapla*, *M. incognita*, *M. incognita acrita*, *M. javanica*, *Pratylenchus penetrans*, *Radopholus similis*, *Paratrichodorus christiei*
Catharanthus roseus (vinca)	*Meloidogyne* sp., *Aphelenchoides* sp., *Ditylenchus dipsaci*
Clerodendrum sp. (bleeding heart vine)	*Meloidogyne arenaria*, *M. incognita*
Crossandra infundibuliformis	*Aphelenchoides fragariae*
Cyclamen persicum	*Aphelenchoides fragariae*, *A. ritzemabosi*, *Ditylenchus dipsaci*, *Meloidogyne incognita*, *M. incognita acrita*, *Pratylenchus* sp.
Dahlia hybrids	*Aphelenchoides besseyi*, *A. fragariae*, *A. ritzemabosi*, *Ditylenchus destructor*, *Paralongidorus maximus*, *Meloidogyne arenaria*, *M. hapla*, *M. incognita*, *M. incognita acrita*, *M. javanica*, *Pratylenchus coffeae*, *Paratrichodorus pachydermus*
Euphorbia pulcherrima (poinsettia)	*Meloidogyne javanica*, *Rotylenchus reniformis*
Fuchsia × *hybrida*	*Meloidogyne hapla*
Gardenia augusta	*Aphelenchoides fragariae*, *Helicotylenchus nannus*, *Meloidogyne arenaria*, *M. hapla*, *M. incognita*, *M. incognita acrita*, *M. thamesi*, *Radopholus similis*, *Rotylenchus reniformis*
Gerbera jamesonii (African daisy)	*Aphelenchoides fragariae*, *Belonolaimus longicaudatus*, *Meloidogyne arenaria*, *M. hapla*, *M. incognita*, *M. incognita acrita*, *M. javanica*, *Paratrichodorus minor*
Hibiscus rosa-sinensis	*Aphelenchoides fragariae*, *Meloidogyne incognita acrita*, *M. javanica*, *Radopholus similis*
Hydrangea macrophylla	*Aphelenchoides besseyi*, *Ditylenchus dipsaci*, *Meloidogyne* sp., *Pratylenchus* sp.
Impatiens wallerana	*Meloidogyne arenaria*
Lantana sp.	*Aphelenchoides fragariae*, *Meloidogyne javanica*, *M. incognita*, *Pratylenchus* sp.
Mimulus sp.	*Aphelenchoides ritzemabosi*
Pelargonium sp. (geranium)	*Aphelenchoides coffeae*, *A. fragariae*, *Meloidogyne arenaria*, *M. hapla*, *M. incognita*, *M. javanica*, *Pratylenchus pratensis*, *Tylenchorhynchus dubius*, *Xiphinema americanum*
Pericallis × *hybrida* (cineraria)	*Aphelenchoides ritzemabosi*, *Meloidogyne javanica*
Primula sp. (primrose)	*Aphelenchoides fragariae*, *A. ritzemabosi*, *Ditylenchus dipsaci*, *Meloidogyne hapla*, *M. javanica*, *M. arenaria*, *Pratylenchus* sp.
Ranunculus sp.	*Aphelenchoides fragariae*, *A. ritzemabosi*, *Ditylenchus dipsaci*, *Meloidogyne hapla*, *Pratylenchus penetrans*
Saintpaulia ionantha (African violet)	*Aphelenchoides fragariae*, *A. ritzemabosi*, *Criconema*, *Meloidogyne arenaria*, *M. arenaria thamesi*, *M. hapla*, *M. incognita*, *M. javanica*, *Paratylenchus projectus*, *Pratylenchus pratensis*, *Rotylenchus brachyurus*, *Scutellonema brachyurum*
Schizanthus sp. (butterfly flower)	*Meloidogyne javanica*
Schlumbergera truncata (Thanksgiving cactus)	*Pratylenchus vulnus*
Sinningia speciosa (gloxinia)	*Aphelenchoides fragariae*, *A. ritzemabosi*, *Meloidogyne arenaria*, *M. javanica*
Solanum pseudocapsicum (Jerusalem cherry)	*Heterodera tabacum*, *Meloidogyne arenaria*, *M. incognita acrita*, *M. javanica*

Selected References

Dropkin, V. H. 1989. Introduction to Plant Nematology. 2nd ed. John Wiley & Sons, New York.

Goodey, J. B., Franklin, M. T., and Hooper, D. J. 1965. The Nematode Parasites of Plants Catalogued under their Hosts. Commonwealth Agricultural Bureaux, Kew, England.

Mai, W. F., and Lyon, H. H. 1975. Pictorial Key to Genera of Plant Parasitic Nematodes. 4th ed. Cornell University Press, Ithaca, NY.

Root-Knot Nematodes

Root-knot nematodes cause considerable economic losses on food, fiber, and ornamental plants worldwide. The common name is derived from the knotlike growths produced on roots. Although root-knot nematodes are potentially destructive, their relatively long generation time (about 30 days) and their infrequent association with potted plants cause them to be only a minor problem for propagators of flowering potted plants.

Symptoms

The diagnosis of root knot is facilitated by the striking symptoms the nematodes produce on roots. Galls range in size from little more than root swellings to large, clubby structures 1 cm or more in diameter (Fig. 56). Symptoms are dependent on the extent of infestation, the host, and the species of nematode. *Meloidogyne hapla* produces galls that are relatively small and have radiating lateral roots, distinguishing them from galls produced by the other three common species. *M. arenaria* also produces small, beadlike galls but without the lateral roots. *M. incognita* and *M. javanica* may produce discrete galls, but these galls often coalesce into large, clubby structures. In African violets (Plate 218), the crown may become swollen and galls may form at the bases of the petioles. Aboveground symptoms include dwarfing, yellowing, and wilting, but these symptoms are not unique to nematodes. In some cases, plants infected with root-knot nematodes have increased height but less mass.

Causal Organism

The genus *Meloidogyne* Goeldi is represented by approximately 55 species, the most commonly encountered of which are *M. incognita* (Kofoid & White) Chitwood, *M. javanica* (Treub) Chitwood, *M. hapla* Chitwood, and *M. arenaria* (Neal) Chitwood. Root-knot nematodes are characterized by sexual dimorphism. Females are distinctively globose (440–1,300 × 300–700 µm), and males and juveniles are vermiform. In addition to morphological features, these four species (including six races) can be identified by differential host response.

Host Range and Epidemiology

Meloidogyne has a wide host range. Nearly all of the plants covered in this compendium are susceptible (Table 18). Worldwide distribution of infected plant material has resulted in the common occurrence of these nematode species in crop plants. *M. hapla,* the northern root-knot nematode, is distributed in the cooler climates of northern North America, northern Europe, and northern Asia. *M. incognita, M. javanica,* and *M. arenaria* are warm-climate species. Because of the considerable movement of potted plants and the moderate temperatures in the greenhouse, any root-knot species could be encountered, regardless of the geographic region.

Meloidogyne is a sedentary endoparasite. Females produce egg masses containing up to 1,000 eggs extruded in a gelatinous matrix. The first instar juveniles undergo one molt in the egg. Molting and hatching are stimulated by plant root exudates. The resulting second-stage juveniles emerge, enter the root, and position themselves next to or within the vascular cylinder, where their presence results in the formation of giant cells and swelling of the root. Two more molts occur within the root. The female develops into the characteristic globose adult and is positioned in the gall so that eggs will be extruded outside the root. The male also swells but becomes vermiform after the last molt and leaves the root.

In the presence of a susceptible host, soil temperature and moisture are the two most important environmental conditions affecting the life cycle. In the greenhouse, both of these factors are often ideal for nematode reproduction. *M. hapla* can invade roots at temperatures of 5–35°C (optimum 15–20°C). Growth and reproduction occur at 15–30°C (optimum 20–25°C). Temperature ranges for *M. javanica* and related species are approximately 5°C higher than for *M. hapla.*

Management

Management practices include the use of soilless growing media, steam pasteurization or chemical treatment of field soil, disease-free plant material, genetic resistance, and avoidance of contamination. If plants are to be rooted in nursery soil outdoors, the application of nematicides or fumigants may be necessary. Steam is the safest and most effective way to treat soil to be used in pots or in raised beds in the greenhouse. Soilless media is free of plant-parasitic nematodes. Rhizomes, corms, or bulbs used as propagative material may harbor nematodes.

Cultivars of hibiscus (*Hibiscus rosa-sinensis* L.) vary in their susceptibility to *M. javanica* and *M. incognita;* in contrast, cultivars of *Impatiens wallerana* Hook. f. have generally been found to be susceptible to *M. arenaria.* Cultivar selection may be important for the end user if the plants are set outdoors in nematode-infested soil.

Nematodes are killed at relatively low temperatures (about 50°C). Soil temperatures of 70°C for 30 min will control other pests as well, so this higher temperature is preferable for soil treatment. Soil fumigation with metham sodium, 1,3-dichloropropene, or a mixture of methyl bromide and chloropicrin prior to planting is also effective for nematode control. Oxamyl is registered in the United States for postplant application for control of some species of root-knot nematodes on certain potted plants.

Selected References

Benson, D. M., and Barker, K. R. 1982. Susceptibility of Japanese boxwood, dwarf gardenia, Compacta (Japanese) holly, Spiny Greek and Blue Rug junipers, and nandina to four nematode species. Plant Dis. 66:1176-1179.

Eisenback, J. D., Hirschmann, H., Sasser, N. N., and Triantaphyllou, A. C. 1981. A Guide to the Four Most Common Species of Root-Knot Nematodes (*Meloidogyne* spp.) with a Pictorial Key. North Carolina State University and the United States Agency for International Development, Raleigh.

Hartman, K. M., and Sasser, J. N. 1985. Identification of *Meloidogyne* species on the basis of differential host test and perineal-pattern morphology. Pages 69-77 in: An Advanced Treatise on *Meloidogyne*. Vol. 2, Methodology. K. Barker, C. Carter, and J. Sasser, eds. North Carolina State University, Raleigh.

Lehman, P. S. 1984. Nematodes which cause decline of gardenia. Fla. Dep. Agric. Div. Plant Ind. Nematol. Circ. 111.

McSorley, R., and Marlatt, R. B. 1983. Reaction of *Hibiscus rosa-sinensis* cultivars to two species of root knot nematodes. HortScience 18:85-86.

Taylor, A. L., and Sasser, J. N. 1978. Biology, Identification and Control of Root-Knot Nematodes. North Carolina State University and the United States Agency for International Development, Raleigh.

Walker, J. T. 1980. Susceptibility of Impatiens cultivars to root-knot nematode, *Meloidogyne arenaria*. Plant Dis. 64:184-185.

Foliar Nematodes

As their common name implies, foliar nematodes cause disease by feeding directly on foliage. No threshold level for these nematodes is tolerable in potted plants because their generation time is relatively short and they destroy the aesthetic value of the plant. They are not common in potted plants, but they can be an important and persistent problem when they do occur.

Symptoms

Depending on the host, *Aphelenchoides* spp. feed ectoparasitically or endoparasitically, resulting in a variety of symptoms. Endoparasitic feeding on foliage often results in chlorotic or brown to purple or black, water-soaked lesions, which are angular in shape because of delimitation by the veins of the leaf (Fig. 57 and Plate 219). Lesions are not angular in *Begonia* spp., and bronzing, yellowing, or reddening of the foliage occurs (Plates 220 and 221). The hue is dependent on light intensity, cultivar, and age of the lesion. Colonized leaves eventually desiccate but remain attached to the plant. On begonia, *A. fragariae* and the bacterium *Xanthomonas campestris* pv. *begoniae* (Takimoto) Dye can occur together, resulting in much greater damage than either of the pathogens can cause alone.

On African violets (*Saintpaulia ionantha* Wendl.), ectoparasitic feeding may occur on immature leaves and flower buds. The lower surfaces of leaves develop small, sunken lesions, which eventually extend through to the top surfaces (Plate 222). Young leaves at the center of the plant become cupped and distorted, and the entire plant may be dwarfed (Fig. 58). The injury may be confused with that caused by the cyclamen mite, *Phytonemus pallidus* Banks.

Causal Organisms

The most common foliar nematodes (also known as bud and leaf nematodes) are *A. fragariae* (Ritzema Bos) Christie, *A. ritzemabosi* (Schwartz) Steiner & Buhrer, and *A. besseyi* Christie. *Aphelenchoides* spp. are vermiform, 0.45–1.20 mm in length, and possess a large metacorpus, a distinguishing feature that occupies three-fourths or more of the width of the body (Fig. 55). They also have a delicate stylet. The genus contains more than 227 species, many of which are mycophagous and some of which are marine inhabitants. *A. fragariae*, *A. ritzemabosi*, and *A. besseyi* can be distinguished by morphological characteristics.

Host Range and Epidemiology

A. fragariae, named for its role in causing spring dwarf disease of strawberry (*Fragaria* × *ananassa* Duchesne), attacks more than 250 plants in 47 families. Flowering potted plant hosts are listed in Table 18. *A. fragariae* occurs in temperate regions and is distributed from Massachusetts to Florida in the United States. In greenhouses, it would be expected to occur in any region on a susceptible host. *A. ritzemabosi* has been reported on 190 species of plants throughout the temperate region and is more likely to be found on plants in the Compositae than *A. fragariae*. *A. fragariae* is common on ferns, whereas *A. ritzemabosi* is not. *A. besseyi* is not as common as the other two species but has been reported on *Dahlia* hybrids and *Hydrangea macrophylla* (Thunb.) Ser.

Foliar nematodes can survive in soil for only several months. Infected but asymptomatic plants are probably the major source of introduction into flower crops. Adult foliar nematodes easily survive desiccation within plant tissues, remaining viable in dead leaves for up to 3 years. Foliar nematodes migrate up the plant when a continuous film of moisture is available. Upward migration is impeded when water is moving down, such as during rain or irrigation. Ingress into the foliage can occur by direct penetration or by

Fig. 57. Wedge-shaped areas of necrosis in columbine leaves infested with foliar nematodes (*Aphelenchoides* sp.).

Fig. 58. Foliar nematode (*Aphelenchoides* sp.) infestation typically causes stunting and hardening of young leaves at the center of African violet plants, often resulting in irregularly shaped plants. (Courtesy R. K. Jones)

movement through the stomates. Reproduction occurs within the infested leaves. As many as 15,000 nematodes have been reported from a single leaf. Foliar nematodes lay 25–32 eggs at a time. The eggs hatch within 3–4 days, and the nematodes mature in 6–12 days. The entire life cycle may be completed within 14 days, allowing very rapid population increase.

Management

Avoidance of infected plant material is the most important method of controlling foliar nematodes. Splashing irrigation water easily spreads them within a susceptible crop or from one infected crop to another. Vegetative propagation from infected plants will further spread the nematodes. Common hosts such as ferns, *Begonia* spp., and African violet should be routinely inspected, especially when new plants are brought into the greenhouse. However, low levels of infection cannot be detected by inspection. Plants with symptoms of foliar nematodes should be discarded. Hot water treatment for 5 min at 50°C will rid infected tissues of the nematodes, although some plants may be injured. Hot water treatment for 10 min at 46°C has been reported to control both *A. ritzemabosi* and *A. fragariae* on begonia. Weed control within greenhouses and adjacent areas is important since foliar nematodes can infect a number of common weeds. Resistance to foliar nematodes has been noted, but it is not a practical management tool. New cultivars are introduced continually, but screening for resistance to nematodes is not routinely carried out. There are no chemicals registered for postplant control of foliar nematodes in the United States.

Selected References

Allen, M. W. 1952. Taxonomic status of the bud and leaf nematodes related to *Aphelenchoides fragariae*. Proc. Helminth. Soc. Wash. 19:108-120.

Crossman, L., and Christie, T. R. 1936. A list of plants attacked by the leaf nematode (*Aphelenchoides fragariae*). Plant Dis. Rep. 20:155-165.

Franklin, M. T. 1965. *Aphelenchoides*. Pages 131-141 in: Plant Nematology. Tech. Bull. 7. J. F. Southey, ed. Her Majesty's Stationery Office, London.

Franklin, M. T., and Siddiqi, M. R. 1972. *Aphelenchoides besseyi*. Commonwealth Institute of Helminthology Descriptions of Plant-Parasitic Nematodes, no. 4. Commonwealth Agricultural Bureaux, Kew, England.

Gill, D. L. 1934. A leaf nematode disease of begonia. (Abstr.) Phytopathology 24:9.

Lehman, P. S. 1989. A disease of begonia caused by a foliar nematode, *Aphelenchoides fragariae*. Fla. Dep. Agric. Consumer Serv. Div. Plant Ind. Nematol. Circ. 164.

Lehman, P. S. 1990. A disease of African violet caused by foliar nematodes. Fla. Dep. Agric. Consumer Serv. Div. Plant Ind. Nematol. Circ. 180.

Riedel, R. M., and Larsen, P. O. 1974. Interrelationships of *Aphelenchoides fragariae* and *Xanthomonas begoniae* on Rieger begonia. J. Nematol. 4:215-216.

Riedel, R. M., Pieron, D. Q., and Powell, C. C. 1973. Chemical control of foliar nematodes (*Aphelenchoides fragariae*) on Rieger begonia. Plant Dis. Rep. 57:603-605.

Siddiqi, M. R. 1974. *Aphelenchoides ritzemabosi*. Commonwealth Institute of Helminthology Descriptions of Plant-Parasitic Nematodes, no. 32. Commonwealth Agricultural Bureaux, Kew, England.

Siddiqi, M. R. 1975. *Aphelenchoides fragariae*. Commonwealth Institute of Helminthology Descriptions of Plant-Parasitic Nematodes, no. 74. Commonwealth Agricultural Bureaux, Kew, England.

Stokes, D. E., and Alfieri, S. A., Jr. 1968. A foliar nematode and a *Phytophthora* parasitic to gloxinia. Proc. Fla. State Hortic. Soc. 81:376-380.

Strider, D. L. 1973. Control of *Aphelenchoides fragariae* of Rieger begonias. Plant Dis. Rep. 57:1015-1019.

Strumpel, H. 1967. Beobachtungen zur lebensweise von *Aphelenchoides fragariae* in Lorraine-begonien. Nematologica 13:67-72.

Stem and Bulb Nematodes and Potato Rot Nematodes

Two species of *Ditylenchus*, *D. dipsaci* (Kühn) Filipjev, the stem and bulb nematode, and *D. destructor* Thorne, the potato rot nematode, damage flowering potted plants. *D. dipsaci*, named after teasel, *Dipsacus fullonum* L., has a wide host range but is best known for its damage to bulb crops such as daffodil (*Narcissus* spp.), tulip (*Tulipa* spp.), hyacinth (*Hyacinthus orientalis* L.), and onion (*Allium cepa* L.). *D. destructor* is important on *Dahlia* spp.

Symptoms

D. dipsaci causes stem necrosis and distortion (swelling and/or twisting) of stems, leaves, and buds. This nematode survives in soil and in weed hosts and attacks bulbs and young shoots. *D. dipsaci* moves intercellularly in young tissues, dissolving the middle lamella and inciting hypertrophy and hyperplasia. Shoots typically become swollen, distorted, and shortened. When dahlia roots are invaded by *D. destructor*, transverse or radial cracking may occur, but this symptom is not unique to nematode injury. A brown to black, dry, granular rot occurs, and subsequent colonization by fungi and bacteria may result in soft rot.

Causal Organism

Ditylenchus spp. are vermiform, long, and slender (0.8–1.4 mm) and possess an inconspicuous stylet. The head region is slightly set off from the body, the stylet is 10–12 μm in length, and the tail is distinctly pointed. The body becomes nearly straight when heat killed.

Host Range and Epidemiology

D. dipsaci occurs on *Cyclamen persicum* Mill., *Dahlia* hybrids, *Hydrangea macrophylla* (Thunb.) Ser., *Primula* spp. (primrose), and *Ranunculus* sp. It is possible that it occurs on additional flowering potted plants not reported in the literature. More than 450 other hosts have been reported, including both monocots and dicots. Eleven or more races have been described, and interbreeding with subsequent alteration in host range has been reported. *D. dipsaci* is bisexual and lays 200–500 eggs over an adult life span of 45–75 days. Optimum temperature for development is 15–20°C, and there is little activity below 10°C or above 22°C. The nematodes eventually migrate to the base of the plant, where they mass into clumps of "eelworm wool." In this state, the nematodes can withstand considerable desiccation.

D. destructor attacks more than 70 hosts, including dahlia rootstocks and begonia. Unlike *D. dipsaci*, *D. destructor* does not survive desiccation well and does not form eelworm wool. It probably survives in field soil by colonizing weed hosts.

Management

Management practices include the use of soilless media, steam or chemical treatment of field soil, disease-free plant material, genetic resistance, and avoidance of contamination. Affected plants should be discarded. If dahlias are to be started in nursery soil outdoors, soil fumigation may be necessary. For soil to be used in pots or raised beds in the greenhouse, steam pasteurization is the safest and most effective treatment. Soilless media are free of plant-parasitic nematodes. Hot water treatment of dahlia rhizomes at 44–45°C for 3 hr will free them of *Ditylenchus*.

Selected References

Esser, R. P., and Smart, G. C., Jr. 1977. Potato rot nematode *Ditylenchus destructor* Thorne, 1945. Fla. Dep. Agric. Consumer Serv. Div. Plant Ind. Nematol. Circ. 28.

Goodey, J. B. 1965. *Ditylenchus* and *Anguina*. Pages 47-58 in: Plant Nematology. Tech. Bull. 7. J. F. Southey, ed. Her Majesty's Stationery Office, London.

Greco, N., Vovlas, N., and Inserra, R. N. 1991. The stem and bulb nematode *Ditylenchus dipsaci*. Fla. Dep. Agric. Consumer Serv. Div. Plant Ind. Nematol. Circ. 187.

Hooper, D. J. 1972. *Ditylenchus dipsaci*. Commonwealth Institute of Helminthology Descriptions of Plant-Parasitic Nematodes, no. 14. Commonwealth Agricultural Bureaux, Kew, England.

Hooper, D. J. 1973. *Ditylenchus destructor*. Commonwealth Institute of Helminthology Descriptions of Plant-Parasitic Nematodes, no. 21. Commonwealth Agricultural Bureaux, Kew, England.

Jensen, H. J., Smithson, H. R., and Loring, L. B. 1958. Potato-rot nematode, *Ditylenchus destructor* Thorne 1945, found in *Dahlia* corms. Plant Dis. Rep. 42:1357-1359.

Part II. Noninfectious Diseases and Arthropod Injuries

Air Pollution Injury

Air pollution is not a major constraint of the flowering potted plant industry, but there are several problems that may occur locally or episodically. Locally occurring problems may emanate from a point source of pollution such as an industrial manufacturer or a coal-burning plant.

Ozone is not a point-source pollutant, but its toxicity to both crop plants and natural vegetation is well documented. However, few important examples of ozone injury have been noted in flowering potted plant production. *Begonia*, *Fuchsia* × *hybrida* Hort. ex Vilm., *Dahlia*, and *Pelargonium* spp. (geraniums) are reported to be susceptible to ozone.

Within closed structures such as greenhouses, flower shops, and storage facilities, various gases may at times accumulate and result in plant injury. The most common example of this type of problem is ethylene injury. Ethylene is a naturally occurring plant growth regulator. When ethylene from external sources comes in contact with flowering potted plants, significant losses may occur. Less than 1 ppm of ethylene can affect plants; sensitive plants react to concentrations as low as 0.017 ppm. The most common symptoms include premature loss of flowers, epinasty (a downward bending or twisting of the leaves), and yellowing and abscission of foliage. African violets (*Saintpaulia ionantha* Wendl.) are likely to show petal abscission; poinsettias (*Euphorbia pulcherrima* Willd. ex Klotzsch) show epinasty; and cineraria (*Pericallis* × *hybrida* R. Nordenstam) leaves roll in from the sides (Plate 223). Other symptoms of ethylene injury, apical meristem inhibition and increased axillary branching, are shown by impatiens seedlings (*Impatiens wallerana* Hook. f.) (Plate 224). Calceolaria (*Calceolaria* Herbeohybrida group) and impatiens are sensitive to ethylene; fuchsia and gardenia (*Gardenia augusta* (L.) Merrill) are less sensitive.

The most common source of ethylene is the incomplete combustion of fossil fuels. Improperly constructed or placed smokestacks, downdrafts from ventilating while heaters are still running, cracks in heat exchangers, and exhaust from automotive engines idling outside the greenhouse can all result in crop damage. Unvented space heaters should never be used in a greenhouse. When sprayers powered by internal combustion engines are used, it is important to have good ventilation in the greenhouse. Polyethylene-covered greenhouses are more airtight than glasshouses and thus tend to accumulate polluting gases. Ethylene is also a cause of many postharvest losses. Wrapping plants for shipping, as is done for poinsettia, can result in ethylene production and accumulation around the plant, causing significant loss of plant quality. Storing flowers with fruits that produce ethylene, such as apples, may drastically shorten the life of the flowers.

Sulfur dioxide (SO_2) pollution may also occur in greenhouses. Direct-fired heaters, sulfur pots burned for disease control, CO_2 generators, and industrial emissions are sources of SO_2. Injury is most likely to occur when fuels with high sulfur contents are used. Only high grades of kerosene, propane (e.g., HD5), and other fuels should be used for heating a greenhouse. Acute injury from SO_2 appears as white to brown, dry, scorched, interveinal patches in the middle-aged tissues (Table 19 and Plate 225). There is a distinct demarcation between injured and healthy tissue. SO_2 can cause acute effects at concentrations as low as 0.1 ppm with only 1–7 hr of exposure.

In order to avoid air pollution injury to plants, each heating unit used in a greenhouse must have a smokestack properly vented to the outside (1 square inch of duct per 2,000 Btu) and an adequate oxygen supply directly available to the burner for proper ignition and burning of fuels.

Selected References

Ball, V., ed. 1991. Ball Red Book. 15th ed. George J. Ball, West Chicago, IL.

Bridgen, M. P. 1984. Be aware of gases in your greenhouse. Conn. Greenhouse Newsl. 123:1-2.

Howe, T. K., and Woltz, S. S. 1981. Symptomology and relative

TABLE 19. Susceptibility of Some Flowering Potted Plants to Damage from Exposure to Sulfur Dioxide[a]

Plant	Susceptibility	Symptoms
Begonia Semperflorens-Cultorum hybrids (wax begonia)	Very high	Marginal and interveinal areas become water soaked, then gray brown, and then white to transparent, resembling onion skin; flower margin necrosis
Catharanthus roseus (vinca)	Very low	No damage
Euphorbia pulcherrima (poinsettia)	Low	White, interveinal necrosis (green foliage most susceptible)
Hibiscus rosa-sinensis	Moderate	Leaf chlorosis may occur within 24 hr; marginal and interveinal scorch on leaves, beginning with toothed edge; sepal tip burn
Impatiens wallerana	Low	White, interveinal and marginal necrosis
Pelargonium × *hortorum* (geranium)	High	Sepal tip burn; interveinal and marginal foliar scorch
Verbena × *hybrida*	Very high	Beige to white scorch along leaf edges and interveinally

[a] Adapted from Howe and Woltz, 1981.

susceptibility of various ornamental plants to acute airborne sulfur dioxide exposure. Proc. Fla. State Hortic. Soc. 94:121-123.

Mastalerz, J. 1977. The Greenhouse Environment. John Wiley & Sons, New York.

Nutrient Disorders

Flowering potted plants are commonly subject to nutrient deficiencies and toxicities. Soilless growing media, water alkalinity, and unbalanced nutritional programs all contribute to the problem. It can be difficult to recognize symptoms of nutrient disorders because they are often subtle or not unique to the cause. Also, different cultivars may vary widely in their susceptibility to deficiencies or toxicities. Tissue analysis is most useful for diagnosis; however, information regarding desirable nutrient tissue levels is lacking for many of the flowering potted plants.

Deficiencies

Nutrient deficiencies cause chronic disease in plants. When nutrients are lacking, important biomolecules like chlorophyll, DNA, RNA, proteins, and lipids cannot be manufactured. Enzymes may not carry out important chemical transformaions. In general, plant growth is slowed, and susceptibility to disease may increase. Flowering potted plants may be dwarfed, develop chlorosis or necrosis, have fewer flowers, and otherwise be unattractive. Of the macronutrients, phosphorus, potassium, and sulfur are not commonly found to be deficient in potted plants. Deficiencies of nitrogen and, in some cases, calcium occur more commonly. Of the micronutrients, iron and molybdenum are occasionally found to be deficient.

Nitrogen is easily leached and must be supplied to plants frequently to prevent deficiency. Nitrogen deficiency results in chlorosis, which is first apparent at the margins of lower leaves and then progresses to the entire plant. Leaf size may be reduced and internodes shortened. Nitrogen levels in newly expanded leaves should be 2.5–4.5% of dry weight.

Calcium deficiency does not commonly occur, but some cultivars of poinsettia (Euphorbia pulcherrima Willd. ex Klotzsch), such as Gutbier V-14 Glory and Celebrate II, develop bract necrosis when foliar tissue calcium levels are low, a condition that is difficult to distinguish from bract necrosis caused by Botrytis (Plate 17). High humidity, frequent irrigation, and high levels of ammonium nitrogen during bract formation may decrease calcium uptake by the plant and cause bract necrosis. Calcium sprays appear to be a more efficient way of preventing bract necrosis than provision of calcium to the growing medium. Reports of normal or optimum levels of foliar calcium vary from 0.7 to 2.0%.

Iron can become deficient when the pH of the growing medium is above 7. However, the occurrence of iron deficiency is largely dependent on the specific requirements of the plant. Chlorosis of the youngest leaves, often with the veins remaining green, is the most common symptom of iron deficiency (Plate 226). Foliar tissue levels should be 75–125 $\mu g/g$.

Magnesium deficiency in poinsettia is evidenced by interveinal chlorosis of the lower leaves. Severely affected leaves develop marginal necrosis and show a downward curling of the margins. Foliar magnesium levels of less than 0.22% will result in symptoms on the lower leaves. Excessive potassium in the soil can induce a magnesium deficiency.

Molybdenum is needed in small amounts by plants, but the use of soilless media and fertilizers lacking this element can result in deficiencies in poinsettia. Symptoms are similar to those of nitrogen or iron deficiency and ammonium toxicity. Plants may be stunted, leaves are small and chlorotic, and leaf margins may become scorched (Plate 227). Leaves tend to curl upward. For poinsettia, foliar levels of molybdenum vary from cultivar to cultivar. Molybdenum levels of less than 0.3–0.5 $\mu g/g$ indicate a deficiency. A high foliar level of nitrate nitrogen (6,000–14,000 $\mu g/ml$) also indicates that molybdenum is deficient.

Toxicities

Excessive levels of nutrients may be toxic to some plants. The most common example is soluble salts toxicity. This occurs when too high a concentration of fertilizer salts is present in the soil solution. High salts may occur from miscalculation of dilutions, poorly calibrated or malfunctioning fertilizer injectors, or simply poor judgment by the grower. Soluble salts toxicity is usually evidenced by chlorosis or necrosis of the leaf margins that begins with the lower leaves (Plate 228). In some plants, leaves become dark green and growth may be stunted. In severe cases, wilt or defoliation may occur (Plates 229 and 230). For soilless growing media, conductivity levels in excess of 300 $dS \cdot m^{-1}$, as measured by a 1:2 dilution of soil and water, are usually an indication of excessive fertilizer salts. Plants vary in their tolerance to high salt levels, and their stage of growth can also be important. New Guinea impatiens (Impatiens hybrids) that are less than 6 weeks old are adversely affected at soil conductivity levels greater than 150 $dS \cdot m^{-1}$. For field soil, the figure is about half that for soilless media. Soluble salts are readily leached from the growing medium.

The macronutrient elements (nitrogen, phosphorus, potassium, sulfur, calcium, and magnesium) are generally not toxic in high amounts. However, high levels of calcium or potassium may interfere with root uptake of magnesium. Foliar calcium levels of 2.9% were reported to result in a significant reduction in total bract area of poinsettia. Excessive rates of nitrogen may result in an increase in susceptibility to bacterial soft rot and other diseases.

Excessive levels of ammonium (NH_4) in the growing medium may interfere with the uptake of calcium, but more commonly, ammonium is directly toxic to plants. Excessive levels can cause reduced growth, interveinal chlorosis, foliar marginal chlorosis or necrosis, and damage to the root system (Plate 231). Soil ammonium levels as measured by the saturated paste method should not exceed 12 ppm, while more than 25 ppm is considered injurious when measurement is done with a 1:2 dilution of soil and water. Fertilizer programs should provide no more than one-half the total nitrogen in the ammonium and/or urea form for flowering potted plants. Ammonium toxicity may occur when fertilizers containing urea or ammonium sulfate are used. Excessive levels of ammonium may also occur after steaming of organic soils, especially those containing manure. The conversion of ammonium to nitrate is carried out by soil microorganisms that are nonexistent or in low numbers in soilless growing media. The conversion can be inhibited by certain pesticides; cool, wet soil; low pH; excessive soluble salts; and poor aeration. Fertilizers containing significant levels of ammonium should be avoided when a soilless medium is used. Ammonium is difficult to leach from the growing medium.

Micronutrients are generally toxic when present in high amounts. In particular, boron, manganese, aluminum, and iron can cause various problems to plants. Plants can vary considerably from cultivar to cultivar in their susceptibility to nutrient toxicities. The pH of the growing medium can significantly affect the plant uptake of micronutrients.

Manganese and iron toxicity is relatively common in greenhouse crops when the pH of the soil is low and additional micronutrients are being added to the crop from fertilizer sources. Steaming soil may also release manganese into the soil solution. When foliar levels of manganese and iron are high, the upper leaves become chlorotic, especially at the margins. Small dots of necrosis may occur scattered on the margins of lower leaves. Chlorosis is also common. Margins may become necrotic. Some geranium cultivars (*Pelargonium* spp.) develop leaf-edge burn and necrotic flecking when grown at a pH of less than 6.0 (Plates 232 and 233). Foliar levels greater than 800 µg/g in chrysanthemum (*Dendranthema* × *grandiflorum* Kitam.) indicate excessive levels of manganese.

Excessive amounts of boron can cause scorching of the leaf tips and margins of some plants. Foliar levels greater than 100 µg/g indicate excessive levels of boron.

Selected References

Ball, V., ed. 1991. Ball Red Book. 15th ed. George J. Ball, West Chicago, IL.

Biernbaum, J. A., Carlson, W. H., Schoemaker, C., and Heins, R. D. 1988. Low pH causes iron and manganese toxicity. Greenhouse Grower 6:92-97.

Bould, C., Hewitt, E. J., and Needham, P. 1983. Diagnosis of Mineral Disorders in Plants. Vol. 1, Principles. Her Majesty's Stationery Office, London.

Cox, D. A. 1988. Lime, molybdenum, and cultivar effects on molybdenum deficiency of poinsettia. J. Plant Nutr. 11:589-603.

Cox, D. A. 1992. Poinsettia cultivars differ in their response to molybdenum deficiency. HortScience 27:892-893.

Cox, D. A., and Seeley, J. G. 1980. Magnesium nutrition of poinsettia. HortScience 15:822-823.

Cox, D. A., and Seeley, J. G. 1984. Ammonium injury to poinsettia: Effects of NH4N:NO3-N ratio and pH control in solution culture on growth, N absorption, and N utilization. J. Am. Soc. Hortic. Sci. 109:57-62.

Engelhard, A. W., ed. 1989. Soilborne Plant Pathogens: Management of Diseases with Macro- and Microelements. American Phytopathological Society, St. Paul, MN.

Harbaugh, B. K., and Woltz, S. S. 1989. Fertilization practice and foliar-bract calcium sprays reduce incidence of marginal bract necrosis of poinsettia. HortScience 24:465-468.

Joiner, N. J., Poole, R. T., and Conover, C. A. 1983. Nutrition and fertilization of ornamental greenhouse crops. Hortic. Rev. 5:317-401.

Judd, L. K., and Cox, D. A. 1992. Growth of New Guinea impatiens inhibited by high growth-medium electrical conductivity. HortScience 27:1193-1194.

Knauss, J. F. 1986. The role of boron in plant nutrition. Grower Talks, Jan., pp. 106-108, 110, 112.

Mastalerz, J. 1977. The Greenhouse Environment. John Wiley & Sons, New York.

Winsor, G., and Adams, P. 1987. Diagnosis of Mineral Disorders in Plants. Vol. 3, Glasshouse Crops. Her Majesty's Stationery Office, London.

Edema

Edema, a noninfectious disorder that is frequently mistaken for a contagious disease, refers to raised blisters that occur primarily on the undersides of leaves. The blistered cells may coalesce, burst, and become corky. Edema is particularly common on certain cultivars of ivy geranium, *Pelargonium peltatum* (L.) L'Hér. (Plate 234), and may also be observed on the florist's geranium, *P.* × *hortorum* L. H. Bailey. Lower leaves severely affected with edema may turn yellow, dry, and abscise. Other flowering potted plants susceptible to edema are *Hydrangea, Kalanchoe,* and *Solanum* spp.

Environmental factors are of primary importance in edema development, especially high relative humidity and high soil temperature. Edema may also develop in response to feeding by twospotted spider mites or as a phytotoxic response to sprays, particularly oil-based materials that may clog stomates.

General recommendations for edema control include the use of a well-drained growing medium and avoidance of exces-sively warm conditions during the day. Plants should be well spaced for good air circulation. To minimize edema development on an ivy geranium crop, high nitrogen and iron levels should be maintained in the tissues, pH levels should be 5.0–5.5, and excessive soluble salts in the growing medium should be avoided. In order to promote drainage, saucers should be removed from hanging baskets while plants are in the greenhouse. Light levels are also important. Ivy geraniums should be grown at moderate light intensities (2,500–4,000 footcandles) to reduce edema severity. Some cultivars of ivy geranium will perform best and show the least edema if grown at 2,000–2,500 footcandles. Greenhouse temperatures should be maintained below 30°C and relative humidity between 60 and 70%.

Selected Reference

White, J. W., ed. 1993. Geraniums IV. Ball Publishing, Geneva, IL.

Pesticide and Growth Regulator Injuries

There is very little tolerance in the potted plant industry for diseased, insect-infested, malformed, or otherwise unattractive plants. To consistently produce high-quality plants, growers often utilize pesticides. Unfortunately, pesticide applications may occasionally be phytotoxic; i.e., they may cause significant damage to plants. Injury may occur when the inappropriate pesticide is chosen, the product is applied in excessive amounts, or drift or runoff moves the product to an unintended target that is sensitive. Ambient temperature, humidity, and moisture may influence whether or not a product will be phytotoxic. Occasionally, a particular cultivar will show sensitivity to a pesticide. Pesticide labels usually indicate plants known to be sensitive to a particular product, but all cultivars of flowering potted plants have not been tested for their sensitivity.

Symptoms of pesticide toxicity vary widely and may include stunted growth, root injury, flower and leaf spots, burning of leaf margins, and dimpling or bronzing of foliage and stems (Plate 235). Plants injured by systemic pesticides usually display stunting, chlorosis, or abnormal leaf and flower formation. Contact materials usually cause spotting, scorching, or bronzing of plant tissues.

Symptoms of pesticide toxicity can mimic those caused by infectious diseases, air pollution, or nutrient deficiencies or toxicities. When these problems are being diagnosed, it is important that the complete history of chemical use on the crop be reviewed. Other important considerations are the length of time between pesticide application and symptom development, the pattern of symptom development in the crop, and injury patterns on the plant that would suggest the direction of spray and obvious shielding effects of leaves. When spotting occurs, a crystalline residue on the spots is indicative (but not conclusive) of contact chemical injury. Generally, pesticide injury is apparent within 1–3 days after application. Emulsifiable concentrate formulations contain organic solvents and are more likely to cause phytotoxic responses than wettable powder formulations.

In most cases, it is not possible to test plant tissue to determine whether pesticides were the cause of injury. However, lesions can be examined and cultured to rule out the possibility of fungal or bacterial pathogens, and additional information can be gathered by means of foliar analysis for nutrients.

Growth regulators used to control height and flowering of plants may also have adverse effects. Plants may fail to grow out of a growth-suppressive treatment or in some cases will develop spots or scorched or chlorotic leaf edges (Plate 236). Symptoms of chlorosis may be temporary. Injury from growth regulators occurs when they are applied at too high a rate or to overly sensitive plants or cultivars. It is important to carefully follow all instructions on the product labels.

Herbicides used to control weeds in or near the greenhouse are sometimes inadvertently allowed to contact crop plants. Volatile herbicides can be especially damaging to a greenhouse crop because of the high temperatures that typically occur within these enclosed structures. Symptoms of herbicide injury vary depending on the activity of the material. Photosynthetic inhibitors cause yellowing and/or necrosis (Plate 237), desiccants cause necrosis, and growth regulators cause abnormal growth of the newly developing tissues (Plate 238). Growth regulator injury causes symptoms similar to those caused by ethylene and some viruses.

Other materials applied to crops may also cause injury under certain conditions. Even excessively cold water may harm foliage of gesneriads (Plate 239). Chlorine bleach uptake by poinsettia roots has been associated with the development of black discoloration on poinsettia stems (Plate 240). Chemical injury in ring patterns might be mistaken for virus infection (Plate 235).

Insect and Mite Injuries

Plant injury symptoms caused by arthropod feeding are rarely mistaken for contagious diseases, since insect or mite bodies, cast exoskeletons, frass, or characteristic feeding injury are usually visible to the careful observer. However, some kinds of damage from insects or mites may be confused with symptoms of contagious disease. Common arthropod-caused injuries that can be difficult to distinguish from disease symptoms include the following examples.

The twospotted spider mite, *Tetranychus urticae* Koch, causes a yellowish stippling of foliage followed by yellowing and necrosis of the oldest leaves. Examination of leaf undersides with a hand lens will clearly show the mites with their characteristic dark spots on the abdomen, pale larvae, round eggs, and (in severe cases) webbing. From a distance, however, twospotted spider mite injury to ivy geraniums (Plate 241) might be confused with symptoms of bacterial blight.

Feeding by leafminer larvae (Diptera: Agromyzidae) causes relatively distinctive serpentine patterns, irregular spots, or blotches on foliage (Fig. 59). The transparency of the mined area or the observation of frass or larvae in the mined tissue will easily distinguish leafminer injury from that caused by fungal, bacterial, and nematode infections. The most harmful leafminer on ornamental greenhouse crops is *Liriomyza trifolii* (Burgess).

Thrips (Thysanoptera: Thripidae) feeding may cause small, irregular, whitish patches on leaf or petal tissue (Plate 242). These spots are accompanied by the deposition of characteristic dark globules of frass that help to identify the cause of injury. The insects themselves are very narrow (0.5 mm) and only 2–3 mm long and thus can easily escape notice, especially in the immature (nymphal) stages, which cause a great deal of the feeding injury (Plate 243). Thrips feeding within buds leads to distortion of the leaves or flowers that is apparent only days or weeks after the injury occurs. The primary thrips affecting greenhouse crops in the United State is the western flower thrips, *Frankliniella occidentalis* (Pergande).

Distortion of the new growth of a plant may be the result of one of several causes, including aphid or thrips feeding, virus infection, or herbicide or ethylene injury. Another cause of leaf distortion that is frequently overlooked is the cyclamen mite, *Phytonemus pallidus* (Banks). Because this tarsonemid mite is too small to be seen clearly with a hand lens, it is very difficult to diagnose as the cause of leaf distortion. Cyclamen mites have elongated (less than 0.3 mm), pale bodies and feed hidden within leaf or flower buds. Their feeding typically causes stunted, hardened new growth, and this injury is easily mistaken for a nutritional disorder. The cyclamen mite has a wide host range, including *Saintpaulia ionantha* Wendl. (African violet), *Cyclamen persicum* Mill., *Fuchsia × hybrida* Hort. ex Vilm., and *Pelargonium* spp. (geranium). The symptoms of injury on African violet (Fig. 60) are fairly characteristic but can be confused with injury caused by foliar nematodes (Fig. 58). Cyclamen infested with cyclamen mites also show hardened, cupped younger foliage.

Distorted growth on plants may also be caused by feeding of the broad mite, *Polyphagotarsonemus latus* (Banks). Broad mites, like cyclamen mites, are tiny, pale, and inconspicuous and require the aid of a microscope for identification. On *Begonia × hiemalis* Fotsch. (Rieger begonia) and *Begonia* Tuberhybrida hybrids (nonstop begonia), broad mite feeding causes bronzing of the interveinal tissue on the leaf underside and stunting (Plate 244). On *Gerbera jamesonii* H. Bolus ex Adlam (African daisy), broad mite feeding may cause a purplish discoloration (Plate 245). On New Guinea impatiens, leaf edges curl under, resulting in a straplike appearance (Plate 246).

Fungus gnats (*Lycoriella* and *Bradysia* spp., Diptera: Sciaridae) are common greenhouse pests. The black-headed larvae feed on fungi in the soil as well as on roots and cuttings of crop plants. They are especially damaging to cuttings of *P. × hortorum* L. H. Bailey (florist's geranium) and *Euphorbia pulcherrima* Willd. ex Klotsch (poinsettia). They enter the base of the cutting and feed, often providing an infection court for a fungal pathogen such as a *Pythium* species and/or preventing normal rooting (Plate 247). Adults may carry fungal

Fig. 59. Injury on columbine caused by feeding of leafminer larvae.

Fig. 60. Stunting of new growth of African violet caused by cyclamen mite infestation. Note the similarity to injury caused by foliar nematodes (Fig. 58).

pathogen inoculum of *Pythium*, *Verticillium*, or *Thielaviopsis* spp. (and probably other genera) from infested pots or the greenhouse floor to uninfested crop plants.

Shore flies (Diptera: Ephydridae), unlike fungus gnats, do not feed directly on crop plants. However, large populations of these insects will feed and develop on algae in growing media or on the greenhouse bench or floor. The larvae are small, light beige, maggotlike worms without a distinctive head capsule. Shore flies have been shown to transmit both fungal and bacterial pathogens to healthy plants in the greenhouse through the deposition of fecal material containing inoculum.

Sooty mold is a black, nonparasitic fungus growth on the plant surface (Plate 248). Nonparasitic fungi such as *Capnodium* sp. and *Meliola* sp. grow on the honeydew produced by phloem-feeding insects such as whiteflies (Homoptera: Aleyrodidae), aphids (Homoptera: Aphididae), and scales (Homoptera: Diaspidae, Coccidae, and Pseudococcidae). Accumulations of sooty mold reduce the aesthetic quality and marketability of a crop, even though the plants are not directly injured.

Insect pests common to the greenhouse environment may also have an important role as vectors of plant viruses. The western flower thrips is the most important vector of impatiens necrotic spot and tomato spotted wilt tospoviruses. The melon aphid, *Aphis gossypii* Glover, and the green peach aphid, *Myzus persicae* (Sulzer), both very common on flowering potted plants, are capable of vectoring a number of viruses that may affect flower crops. *Bemisia tabaci* (Gennadius), the sweetpotato whitefly, is a vector for many viruses worldwide, but the strain of this insect found in greenhouses (the silverleaf whitefly, or strain B) has not yet been implicated as a vector for any diseases of flowering potted plants.

Selected References

Baker, J. R., ed. 1978. Insect and Related Pests of Flowers and Foliage Plants. North Carolina Agricultural Extension Service, Raleigh.

Powell, C. C., and Lindquist, R. K. 1992. Ball Pest and Disease Manual. Ball Publishing, Geneva, IL.

Glossary

A—acre
C—centigrade or Celsius
cm—centimeter (0.39 in.; 10 mm = 1 cm)
F—Fahrenheit
g—gram
gal—gallon
in.—inch (2.54 cm)
kg—kilogram (2.23 lb)
L—liter (1.06 quarts)
lb—pound (1 lb = 453.59 g)
m—meter (39.37 in.)
mg—milligram
ml—milliliter (1 ml = 0.001 L)
mm—millimeter (0.04 in.; 10 mm = 1 cm)
μg—microgram (10^{-6} g)
μm—micrometer (10^{-6} m)
oz—ounce (28.34 g) or fluid ounce (29.6 ml)
ppm—parts per million

abiotic—nonliving
abscise (n. abscission)—to fall off, as leaves, flowers, fruits, or plant parts do
acervulus (pl. acervuli)—erumpent, saucer-shaped, cushionlike fruiting body of a fungus bearing conidiophores, conidia, and sometimes setae
acropetal—upward from the base toward the apex
aerobic—living, active, or occurring only in the presence of oxygen
anamorph—the asexual form (also called the imperfect state) in the life cycle of a fungus that produces asexual spores (such as conidia) or no spores
antheridium—male sexual organ found in some fungi
anthracnose—a disease characterized by necrotic lesions caused by fungi that produce spores borne in acervuli
antibody—a specific protein produced by an organism in response to an antigen
antigen—a substance that stimulates production of antibodies
antiserum—a serum containing antibodies
aphid—insect (Homoptera) that feeds on juices of many types of plants and may serve as a vector of certain virus diseases of plants
aplerotic—pertaining to an oospore that does not fill the oogonium
apothecium (pl. apothecia)—cuplike or saucerlike ascus-bearing fruiting body (ascocarp)
appressorium (pl. appressoria)—flattened swelling on a germ tube or hypha of a parasitic fungus that adheres to the surface of the host before haustoria penetrate the host cell
ascocarp—sexual fruiting body (ascus-bearing organ) of an ascomycete; i.e., apothecium, perithecium, and cleistothecium
ascomycete—fungus that produces sexual spores (ascospores) within an ascus
ascospore—spore produced within an ascus
ascus (pl. asci)—saclike or clavate cell containing ascospores (typically eight) and borne in an ascocarp
aseptate—without cross walls
asexual—vegetative; without sex organs, gametes, or sexual spores; imperfect
avirulent—unable to cause disease; nonpathogenic

bacterium (pl. bacteria)—minute, prokaryotic organism that usually lacks chlorophyll and exists mostly as a parasite or saprophyte
basidiomycete—fungus that forms sexual spores (basidiospores or sporidia) on a basidium

basidiospore—haploid spore produced on a basidium
basidium (pl. basidia)—short, club-shaped, haploid promycelium produced by a basidiomycetous fungus
binucleate—having two nuclei
bioassay—test involving the response of a living cell or organism to an artificial stimulus
biological control—disease or pest control through counterbalance of microorganisms and other natural components of the environment
blight—sudden, severe, and extensive wilting and/or death of leaves, stems, flowers, or entire plants
blotch—irregularly shaped, usually superficial spot or blot

canker—stem lesion with sharply limited necrosis of the cortical tissue
causal agent—anything (biotic or abiotic) capable of causing a disease
chlamydospore—thick-walled asexual spore formed via modification of a hyphal or conidial cell
chlorosis (adj. chlorotic)—failure of chlorophyll development caused by nutritional disturbance or disease; fading of green plant color to light green, yellow, or white
clavate—club-shaped; narrowing toward the base
coenocytic—nonseptate; having nuclei not separated by cross walls
conidiophore—simple or branched fertile hypha on which conidia are produced
conidium (pl. conidia)—asexual spore borne at the tip or side of a specialized hypha called a conidiophore
cortex (adj. cortical)—tissues between the epidermis and phloem in stems and roots
crown—compacted series of nodes from which shoots and roots arise
cultivar—cultivated variety; group of closely related plants of common origin within a species that differ from other cultivars in certain minor details such as form, color, flower, or fruit
culture-index—culture of stem sections on a defined medium to test for the presence of fungal or bacterial vascular pathogens
cuticle—outer sheath or membrane of a plant

damping-off—rapid, lethal decline of germinating seed or seedlings before or after emergence
deuteromycete—one of a class of fungi with no known sexual stages
dieback—progressive death of leaves, stems, or roots from the tips back
diploid—having a double set of chromosomes per cell ($2n$)
disease—any abnormal or malfunctioning process in the host induced by constant association with one or more causal agents
disease complex—diseases resulting from combined or sequential actions of two or more biotic or abiotic agents
disease cycle—chain of events involved in disease development
disinfest—to kill or inactivate disease organisms on the surfaces of seeds or plant parts or in soil before they can cause infection
dissemination—spread of infectious material (inoculum) from a diseased to a healthy plant by wind, water, people, animals, insects, mites, or machinery
drench—pertaining to a liquid chemical substance poured onto the soil around a plant rather than applied as a foliar spray

edema—swelling symptom caused by excessive moisture; appears as numerous small bumps on lower sides of leaves or on stems
ELISA—*see* enzyme-linked immunosorbent assay
endemic—native to one country or geographic region
enzyme—protein that catalyzes a specific biochemical reaction
enzyme-linked immunosorbent assay (ELISA)—serological procedure often used for virus assay

epidemic—general and serious outbreak of disease

epidemiology—study of disease initiation, development, and spread

epidermis—superficial layer of cells on all plant parts

epinasty—abnormal twisting of stems and downward bending of leaves

eradication—control of disease by eliminating the pathogen after it has become established

ethylene—colorless, odorless gas that hastens the senescence of flowers; often emitted by fruit and foliage or as a result of incomplete combustion of oil or gas in heaters

etiolation—yellowing of tissue and elongation of stems caused by low light

etiology—study of the cause or origin of a disease

exudate—substance that is excreted or discharged; ooze

facultative parasite—organism that is normally self-dependent but that is adaptable to a parasitic mode of life

filamentous or **filiform**—threadlike

flagellum (pl. flagella)—appendage that provides locomotion

fleck—minute spot

foliar—pertaining to leaves

foot cell—basal cell, as of a spore or conidiophore

fruiting body—any complex, spore-bearing fungal structure

fumigant—vapor-active chemical

fungicide (adj. fungicidal)—chemical or physical agent that kills or inhibits the growth of fungi

fungistatic—capable of preventing the growth of a fungus without killing it

fungus (pl. fungi)—organism lacking chlorophyll that reproduces by sexual or asexual spores and not by fission

gall—abnormal swelling or localized outgrowth on a plant, often more or less spherical

germ tube—hypha resulting from an outgrowth of the spore wall and/or cytoplasm

girdle—to encircle

gram-positive or **-negative**—pertaining to bacteria that retain or release the violet dye in the Gram stain procedure

guttation—exudation of water from stomates or hydathodes

haploid—having a single complete set of chromosomes

haustorium (pl. haustoria)—specialized hypha within penetrated host cells that probably functions in food absorption

herbicide—chemical or physical agent that inhibits or kills plants

hilum—scar on a spore at the point of attachment

host—living plant attacked by (or harboring) a living parasite and from which the invader obtains part or all of its nourishment

host range—range of plants attacked by a given pathogen

hyaline—transparent or nearly so; colorless

hybrid (v. hybridize)—offspring of two individuals of different genetic character

hydathode—structure specialized for secretion or exudation of water

hyperplasia (adj. hyperplastic)—excessive, abnormal, usually pathological multiplication of the cells of a tissue or organ

hypersensitive—displaying increased sensitivity, as when host tissue dies at the point of attack by a pathogen so that infection does not spread

hypertrophy—excessive, abnormal, usually pathological enlargement of cells in a tissue or organ

hypha (pl. hyphae)—tubular filament of a fungal thallus or mycelium

immune—not affected by or responsive to disease

imperfect state—*see* anamorph

in vitro—in glass or an artificial environment

inclusion body—structure in a plant cell induced by virus infection that may be used in virus classification

indicator plant—plant that reacts to certain viruses, other pathogens, or environmental factors with specific symptoms; such plants are used to identify pathogens or determine the effects of environmental factors

infect (n. infection)—to enter and establish a parasitic relationship with a host plant

infectious—capable of infecting and spreading from plant to plant

infest—to contaminate, as with organisms

inflorescence—flowering portion of a shoot

inoculate—to place inoculum in an infection court

inoculum—pathogen or its parts responsible for producing disease

insecticide—chemical or physical agent used to control or kill insects

intercalary—formed along or within the hypha, not terminal on hyphal tips

intercellular—between or among cells

internode—area between two adjacent nodes on a stem

interveinal—area between leaf veins

intracellular—within or through a cell or cells

isolate—separated or confined spore or microbial culture

latent—present but not manifested or visible

leach—to wash soluble nutrients down through the soil

leaf spot—self-limiting or localized lesion on a leaf

lesion—well-marked but localized diseased area

macroconidium (pl. macroconidia)—long or large conidium relative to microconidia

manual transmission—spread or introduction of inoculum to infection courts by hand manipulation

mechanical inoculation or **transmission**—spread or introduction of inoculum to infection courts (especially wounds) accompanied by physical disruption of host tissues

meristem (adj. meristematic)—plant tissue that functions principally in cell division and differentiation

microconidium (pl. microconidia)—small conidium relative to macroconidia; microspore

microflora—composite of microscopic plants

microscopic—too small to be seen without a microscope

midrib—central, thickened vein of leaves

MLO—mycoplasmalike organism; phytoplasma

mold—any profuse fungal growth

morphology—study of form and structure

mosaic—leaf symptom characterized by mottling or variegated patterns of dark and light green to yellow

motile—able to move

mottle—disease symptom characterized by light and dark areas in an irregular pattern

multinucleate—having more than one nucleus

multiple infection—invasion by more than one parasite

multiseptate—having many septa

muriform—having transverse and longitudinal septa

mutation—abrupt, heritable change in an individual

mycelium (pl. mycelia)—mass of hyphae that comprises the thallus or body of a fungus

necrosis (adj. necrotic)—death, usually accompanied by darkening or discoloration

nematicide—chemical or physical agent that kills or inhibits nematodes

nematode—small, wormlike animal parasitic in plants or animals or free-living in soil or water

nonseptate—without cross walls

obligate parasite—organism that can survive only on or in living tissue and that has not been cultured on laboratory media

oogonium—female gametangium of some oomycetes containing one or more gametes

oomycete (adj. oomycetous)—a fungus that produces oospores

oospore—thick-walled, sexually or asexually derived resting spore of phycomycetous fungi

overwinter—to survive the winter

parasite (adj. parasitic)—an organism living in or on another living organism (host) and obtaining food from it

pasteurization—method of destroying a selective microbial population by heating to a prescribed temperature for a specified period of time

pathogen—organism or agent that causes disease in another organism

pathogenicity—ability to cause disease

pathovar (pv.)—a strain or set of strains with the same or similar characteristics differentiated on the basis of distinctive pathogenicity to one or more plant hosts

perfect state—*see* teleomorph

perithecium (pl. perithecia)—flask-shaped or subglobose, thin-walled fungus fruiting body (ascocarp) containing asci and ascospores

pest—any organism that injures plants or plant products

petiole—the stem of a leaf; the stalk attaching a leaf blade to a stem

pH—measure of acidity (pH 7 is neutral; below pH 7 is acidic; above pH 7 is alkaline)

phialide—conidiophore of fixed length with one or more open ends through which a basipetal succession of conidia develops

phloem—food-conducting tissue in plants

photosynthesis—the manufacture of carbohydrates in the presence of chlorophyll by using light energy

phycomycete (adj. phycomycetous)—one of a group of fungi whose mycelia have no cross walls

phytopathology—plant pathology; the science of plant disease

phytoplasma—prokaryotic organism smaller than conventional bacteria, without rigid cell walls, and variable in shape; previously termed MLO

phytotoxic—harmful to plants

polar—at the ends or poles

primary inoculum—inoculum that initiates rather than spreads or magnifies disease

propagation—reproduction by sexual or vegetative (asexual) means

propagule—any part of an organism capable of independent growth

pseudomonad—bacterium of the genus *Pseudomonas*

pycnidium (pl. pycnidia)—asexual, flask-shaped or globose fungus fruiting body containing conidia (pycnidiospores)

pycnium (pl. pycnia)—haploid, pycnidium-like fruiting body or spermagonium produced by rust fungi

quarantine—legislative control of the transport of plants or plant parts to prevent disease spread

resistance—ability of a host plant to overcome completely or to suppress, prevent, or impede the activity of a pathogen

resting spore—temporarily dormant, usually thick-walled spore

rhizome (adj. rhizomatous)—jointed, underground stem

rhizosphere—microenvironment in soil influenced by plant roots

ring spot—disease symptom characterized by yellowish or necrotic rings with green tissue inside the rings; usually caused by a virus

rogue—to remove and destroy by hand individual plants that are diseased, infested by insects, or otherwise undesirable

rot—softening and disintegration of succulent plant tissue as a result of fungal or bacterial infection

sanitation—destruction of infected and infested plants or plant parts

saprophyte (adj. saprophytic)—an organism that uses nonliving organic matter as food

sclerotium (pl. sclerotia)—hard, usually darkened and rounded mass of dormant hyphae with differentiated rind and medulla

secondary inoculum—inoculum resulting from primary infections

sedentary—remaining in a fixed location

senesce—to decline as with maturation, age, or disease stress

septate—having cross walls

septum (pl. septa, adj. septate)—cross wall

serology—study, detection, and identification of antigens, antibodies, and their reactions

seta (pl. setae)—stiff, hairlike appendage, usually dark and thick walled

shot hole—symptom in which small, round fragments drop out of leaves

sign—indication of disease from direct visibility of the pathogen or its part

sooty mold—dark fungus usually growing in insect honeydew

sp. (pl. spp.)—species; a genus name followed by sp. means that the particular species is undetermined; spp. after a genus name means that several species are being referred to without being named individually

sporangiophore—differentiated hypha that bears a sporangium

sporangium (pl. sporangia)—flasklike fungal structure whose contents differentiate into asexual spores

spore—one- to many-celled reproductive body in fungi and lower plants

sporodochium (pl. sporodochia)—superficial, cushion-shaped, asexual fruiting body

sporulate—to produce spores

spot—limited, chlorotic or necrotic, circular to oval area on a plant

sterile—infertile; free from contaminant organisms

sterilization—method of destroying all microorganisms by heating to 100°C for 20 min generally using free steam. Conditions for sterilization in an autoclave are usually 121°C and 15–17 lb of pressure

stock—portion of the stem and associated root system onto which a scion is grafted; production planting

stoma or **stomate** (pl. stomata or stomates)—regulated opening in the plant epidermis for passage of gases and water vapor

strain—biotype; race; an organism or group of organisms that differs in minor aspects from other organisms of the same species or variety

stroma (pl. stromata)—compact mass of mycelium (with or without host tissue) that supports fruiting bodies

stylet—pointed, slender structure in the mouth portion of a plant-parasitic nematode

substrate—surface or medium on or in which an organism is living and from which it gets its nourishment

surfactant—monomolecular compound used as a detergent that reduces surface tension and provides spreading action when used with pesticides

susceptible—not immune; lacking resistance; prone to infection

symptom—indication of disease by reaction of the host

syn.—synonym(s)

systemic—pertaining to chemicals or pathogens (or single infections) that spread generally throughout the plant body as opposed to remaining localized

teleomorph—the sexual form (also called the perfect state) of a fungus

teliospore—thick-walled resting spore produced by some fungi, notably rusts and smuts, that germinates to form a basidium

thermal inactivation point—temperature at which virus particles are inactivated or lose their infectivity

tissue analysis—analysis of leaf tissues for major and minor elements

tissue culture—the technique of cultivating cells, tissues, or organs in a sterile, synthetic medium

tolerant—capable of sustaining disease without serious damage

toxicity—capacity of a substance to produce injury

translucent—so clear that light rays may pass through

transmission—spread of virus or other pathogen from plant to plant

transpiration—loss of water vapor from aerial parts of plants

variety—group of closely related plants of common origin within a species that differ from each other in certain minor details

vascular—pertaining to conductive (xylem and phloem) tissue

vector—agent (e.g., insect, mite, animal, or human) able to transmit a pathogen (virus, bacterium, fungus, phytoplasma, nematode)

viability (adj. viable)—state of being alive; ability of seeds, fungus spores, and sclerotia to germinate

viroid—smallest known agent of infectious disease; contains a small bit of RNA but no protein

virulence—degree of pathogenicity; capacity to cause disease

virus—submicroscopic, filterable agent that causes disease, multiplies only in living cells, and contains nucleic acid surrounded by protein

virus index—assay of plant tissues for presence of virus

water-soaked—pertaining to plants or lesions that appear wet, dark, and usually sunken and translucent

wilt—drooping of leaves from lack of water; a vascular disease that interrupts the normal uptake and distribution of water

zoospore—asexually produced fungus spore having cilia or flagella and capable of locomotion in water

Selected References

Agrios, G. N. 1978. Plant Pathology. 2nd ed. Academic Press, New York.

Ainsworth, G. C. 1971. Ainsworth and Bisby's Dictionary of the Fungi. 6th ed. Commonwealth Mycological Institute, Kew, England.

Federation of British Pathologists (Terminology Sub-Committee). 1973. A Guide to the Use of Terms in Plant Pathology. Phytopathol. Pap. 17. Commonwealth Mycological Institute, Kew, England.

Henderson, I. F., and Henderson, W. D. 1963. A Dictionary of Biological Terms. 8th ed. J. H. Kenneth, ed. Oliver & Boyd, Edinburgh.

Roberts, D. A., and Boothroyd, C. W. 1972. Fundamentals of Plant Pathology. W. H. Freeman, San Francisco.

Robinson, R. A. 1969. Disease resistance terminology. Rev. Appl. Mycol. 48:593-606.

Stern, W. T. 1973. Botanical Latin. David and Charles, London.

Strider, D. L. 1985. Diseases of Floral Crops, vol. 2. Praeger Scientific, New York.

Index